THE TIMES

Big Book of ULTIMATE KILLER Su Doku Book 1

360 of the deadliest Su Doku puzzles

Published in 2021 by Times Books

HarperCollins Publishers
1 London Bridge Street
London SE1 9GF

HarperCollins*Publishers*
Macken House, 39/40 Mayor Street
Upper, Dublin 1, D01 C9W8, Ireland

www.collinsdictionary.com

10 9 8 7 6 5 4 3

ISBN 978-0-00-798316-2
Previously published as
978-0-00-732663-1, 978-0-00-736452-7,
978-0-00-744065-8

Layout by Davidson's Publishing Solutions

Printed and bound in the UK
using 100% renewable electricity
at CPI Group (UK) Ltd

A catalogue record for this book is available from the British Library.

If you would like to comment on any aspect of this book, please contact us at the above address or online via email:
puzzles@harpercollins.co.uk
Follow us on Twitter @collinsdict
Facebook.com/CollinsDictionary

FSC
www.fsc.org
MIX
Paper | Supporting responsible forestry
FSC™ C007454

This book contains FSC™ certified paper and other controlled sources to ensure responsible forest management.

For more information visit: www.harpercollins.co.uk/green

Contents

Introduction

Welcome to *Ultimate Killer Su Doku*. Killers are my personal favourites, and this level is the pinnacle!

Almost every move will be difficult at this level, so do not be put off when solving the puzzle seems hard going. The techniques described here are sufficient to solve all the puzzles. As you work through the book, each new puzzle will present some new challenges and you just need to think about how to refine your application of the techniques to overcome them. Above all, practice makes perfect in Su Doku.

It is also important to note that the puzzles in this book all use the rule that digits cannot be repeated within a cage. Some puzzle designers allow repeated digits, and you need to be aware of this variation in Killers, but *The Times* puzzles do not.

Cage Integration

You will already know that the sum of all the digits in a row, column or region is 45 so, if as in the bottom right region of Fig. 1 (overleaf), all but one of the cells are in cages that add up to 39, the remaining cell G9 must be 6. This concept can be extended to solve a number of cells and get the puzzle started. It is the only way to start a Killer at this level, and it is essential, if the rest of the puzzle is to be solved, to

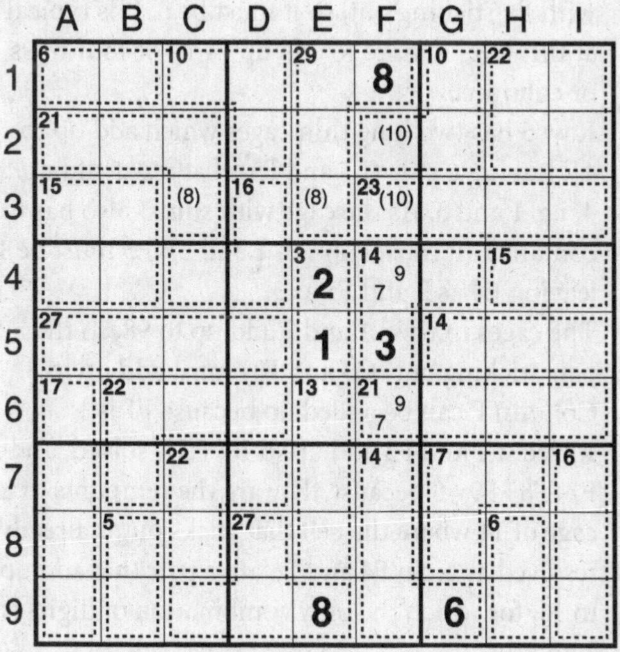

Fig. 1

find every single cell and pair of cells that can be solved in this way before continuing. The others in Fig. 1 are solved as follows:

- In the top right region, the cages add up to 55. Two cells (F2 and F3) are sticking out, and therefore make up a pair that adds up to 10 – indicated by (10) in each.
- The cages in columns A, B, C and D add up to 188

with E9 sticking out, so it must be 8. It is typical at this level to have to add up three or four rows or columns.

- Row 5 has two long thin cages which add up to 41, leaving a pair (E5 and F5) that must add up to 4, i.e. 1 and 3. As the cage with sum 3 also has to contain 1 (it must contain 1 and 2), E5 must be 1, leaving E4 as 2 and F5 as 3.
- The cages in rows 1 and 2 add up to 98, so the pair sticking out (C3 and F3) must add up to 8.
- Column F can be added up because F1 is solved, F2 and F3 add up to 10, F5 is solved, and F7+F8+F9=8 because they are the remnants of a cage of 14 where the cell that sticks out is already resolved as 6. So F4 and F6 are a pair that add up to 16, for which the only combination of digits is 7 and 9.

The integration technique has now resolved quite a lot of the puzzle, probably a lot more than you expected. To get really good at it, practise visualising the shapes made by joining the cages together, to see where they form a contiguous block with just one or two cells either sticking out or indented. Also, practise sticking with this technique and solving as much as possible before allowing yourself to start using the next ones. I cannot emphasise enough how much time will be saved later by investing time in thorough integration. You also need to practise your mental arithmetic so that you can add up the cages quickly and accurately.

Combination Elimination

The main constraint in Killers is that only certain combinations of digits are possible within a cage. Easier puzzles rely on cages where only one combination is possible, e.g. 17 in two cells can only be 8+9. At this level, however, there will be very few easy moves, and most cages will have multiple possible combinations of digits. It is necessary to eliminate the combinations that are impossible due to other constraints in order to identify the one combination that is possible. This concept is used to solve the bulk of the puzzle.

It is useful to identify where the possible combinations all contain the same digit or digits, so you know that digit has to be somewhere in the cage. The digit can then be used for scanning and for elimination elsewhere. It is also useful to identify any digit that is not in any of the combinations and so cannot be in the cage.

The most popular multiple combinations are:

Two cell cages	Three cell cages	Four cell cages
5 = 1+4 or 2+3	8 = 1+2+5 or 1+3+4 (always contains 1)	12 = 1+2+3+6 or 1+2+4+5 (always contains 1+2)
6 = 1+5 or 2+4	22 = 9+8+5 or 9+7+6 (always contains 9)	27 = 9+8+7+3 or 9+8+6+4 (always contains 9+8)
7 = 1+6 or 2+5 or 3+4		28 = 9+8+7+4 or 9+8+6+5 (always contains 9+8)
8 = 1+7 or 2+6 or 3+5		
9 = 1+8 or 2+7 or 3+6 or 4+5		
10 = 1+9 or 2+8 or 3+7 or 4+6		
11 = 2+9 or 3+8 or 4+7 or 5+6		
12 = 3+9 or 4+8 or 5+7		
13 = 4+9 or 5+8 or 6+7		
14 = 5+9 or 6+8		
15 = 6+9 or 7+8		

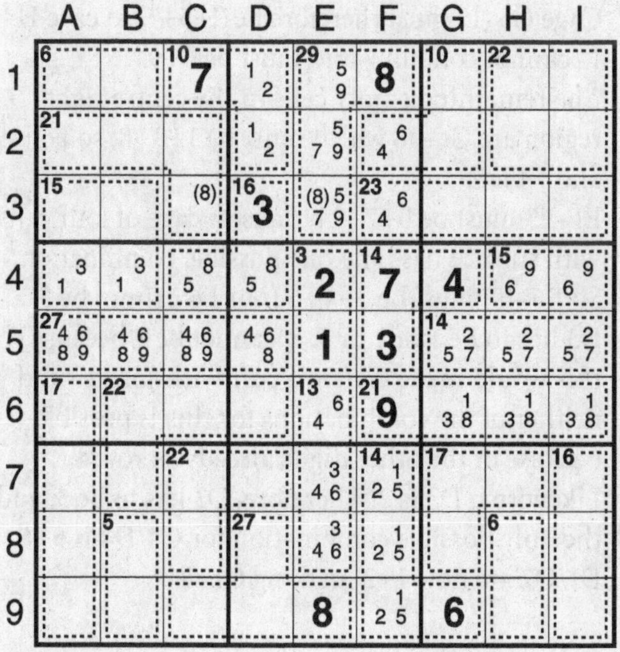

Fig. 2

Working from the last example, the following moves achieve the position in Fig. 2 above:

- F7/F8/F9 form a cage with sum 8; 1+3+4 isn't possible because of the 3 in F5.
- F2/F3 has sum 10 and 4+6 is all that remains in the column.
- Cage A5–D5 cannot be 9+8+7+3 because of the 3 in F5.

- Cage G5–I5 must therefore be 2+5+7, so cage H4/I4 cannot contain 7 and must be 6+9.
- The remaining empty cells in the centre right region are G6–I6 which must be 1+3+8, so F6 is 9 and F4 is 7.
- E1–E3 must be 5+7+9 because a cage of four cells with sum 29 has only one possible combination. So E6–E8 must be 3+4+6 (but E6 cannot be 3).
- D3 has to be 1 or 2 or 3. It cannot be 1 because this would make C4/D4 a pair with sum 15 and neither of the combinations for this is possible because of the other digits already in row 4. Likewise if D3 is 2. Therefore, D3 has to be 3, and the only possible combination for C4/D4 is 5+8.
- D1/D2 are now 1+2, making C1=7.

Getting good at this is rather like learning the times table at school, because you need to learn the combinations off by heart; then you can just look at a cage and the possible combinations will pop into your head, and you can eliminate the ones that are excluded by the presence of other surrounding digits.

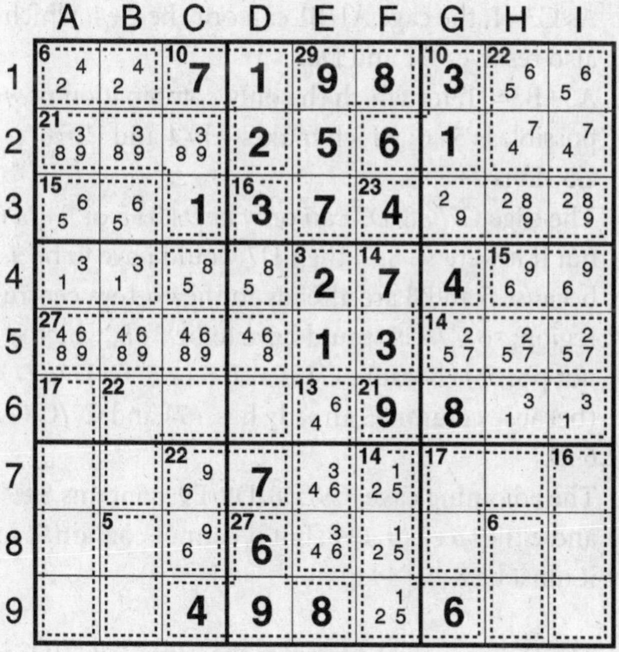

Fig. 3

Further Combination Elimination

To progress from where Fig. 2 finished off to Fig. 3 above:

- Go back to the fact that C3+E3=8, as marked by the (8) symbols; E3 can only be 5, 7 or 9. E3 cannot be 9, as this would require C3 to be –1. It cannot be 5 either, because D3 is 3. So the only possible combination is C3=1 and E3=7.

- As C3=1, the cage A1/B1 can only be 2+4, which also resolves D1 and D2.
- A3+B3=11, for which the only combination now possible is 5+6; which makes F3=4 and F2=6, so G3–I3 is 2+8+9.
- The cage C7/C8/D7 can only be 9+8+5 or 9+7+6, but if it were 9+8+5 then D7 would have to be 9, because 5 and 8 are already in the bottom centre region; so C7/C8 would have to be 5+8, which is not possible because C4 is also 5 or 8. So D7=7 (because column C already has a 7) and C7/C8 is 6+9.
- The adjoining cage, C9/D8/D9/E9, contains 8+9 and either 3+7 or 4+6, but it cannot contain 7, so it must be 8+9+4+6.

The technique used to solve the cage C7/C8/D7 is well worth thinking about, because it is often critical in solving puzzles at this level. We knew that C4 could only be 5 or 8, which means that another pair of cells in the same column, C7/C8, could not be 5+8, because this would require three cells (C4/C7/C8) to contain just two digits.

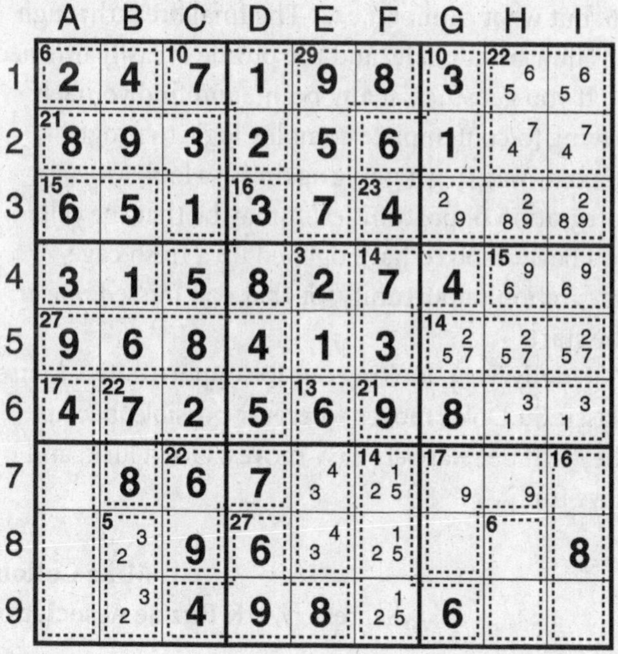

Fig. 4

Inverse Logic

Fig. 4 shows a very interesting situation in the bottom right region – the cage with sum 17 is missing two digits that add up to 8, and so is the cage with sum 16. The possible combinations are 1+7, 2+6 and 3+5, but 2+6 is not possible in either because G9 is already 6, so the 2 must be in the cage H8/H9, making it 2+4. This illustrates another very important technique to think about, the use of inverse logic – not what can go in the cages of 17 and

16, but what cannot, i.e. 2. The final breakthrough has now been made, and the puzzle is easily finished.

If you get stuck at any point, and find yourself having to contemplate complex logic to progress, the best way to get going again is to look for cage integration opportunities. It may be that the cells you have resolved have opened up a fresh cage integration opportunity, or that you missed one at the start.

Finally, keep looking out for opportunities to use classic Su Doku moves wherever possible, because they will be relatively easy moves. Good luck, and have fun.

Mike Colloby
Secretary, UK Puzzle Association

PUZZLES
Book One

1

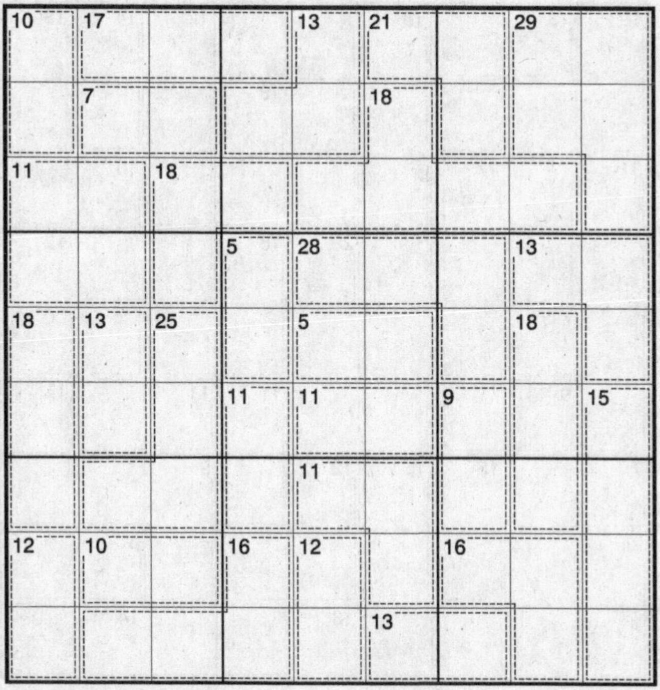

1 HOUR

TIME TAKEN.............................

Deadly

2

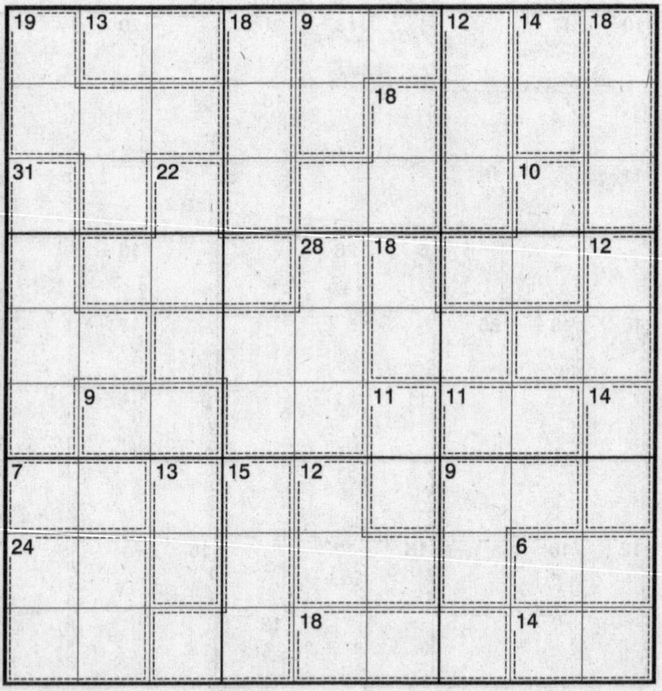

1 HOUR

TIME TAKEN..........................

3

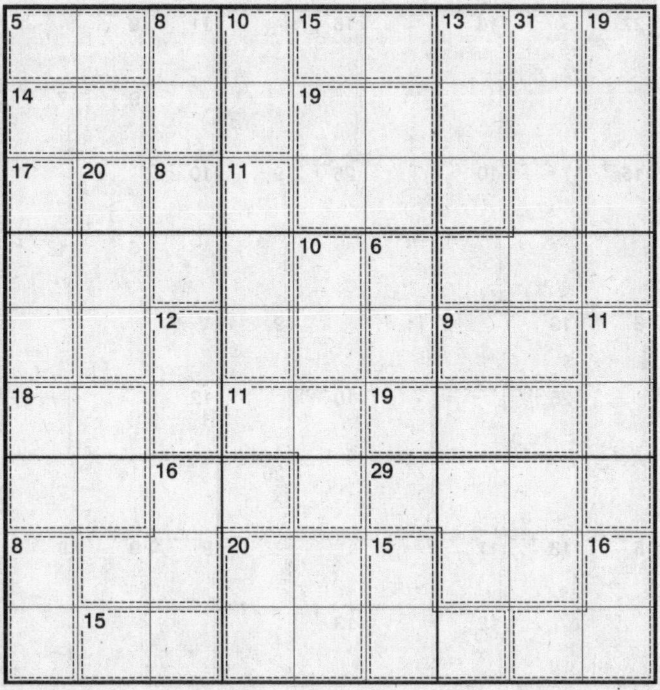

1 HOUR

TIME TAKEN...........................

Deadly

4

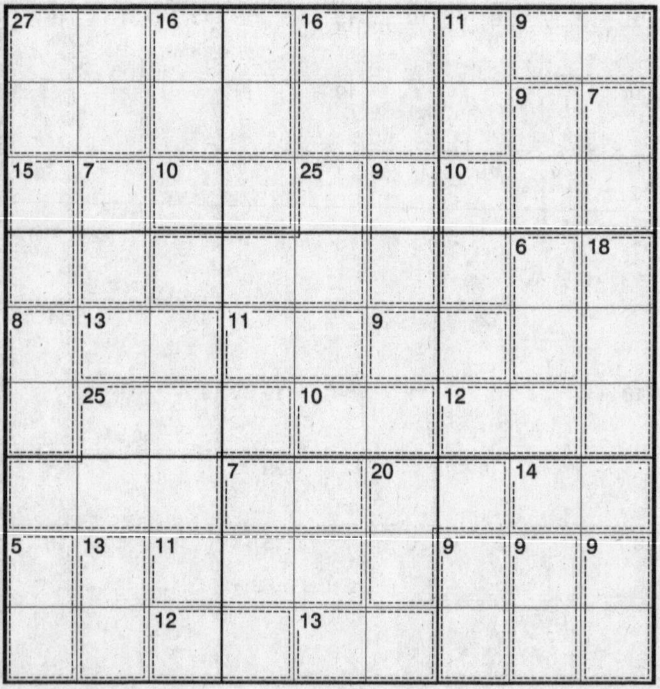

1 HOUR

TIME TAKEN............................

5

1 HOUR 5 MINUTES

TIME TAKEN.............................

Deadly

6

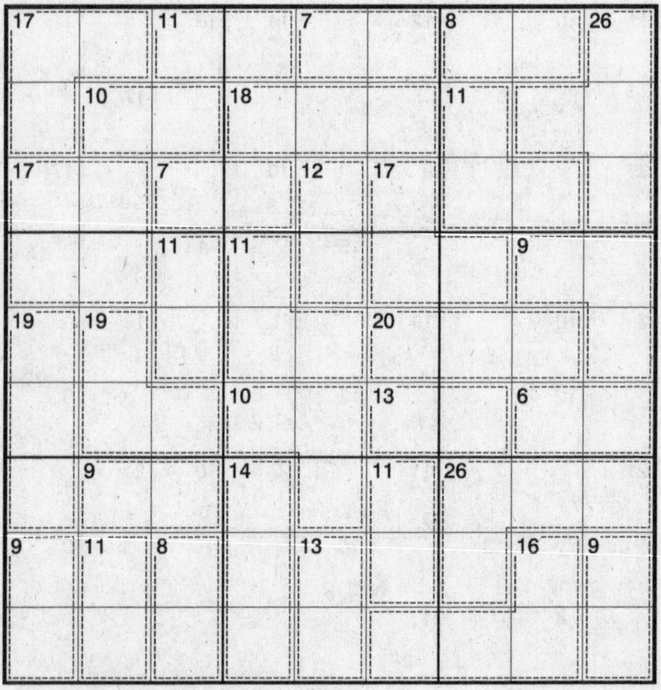

1 HOUR 5 MINUTES

TIME TAKEN............................

1 HOUR 5 MINUTES

TIME TAKEN............................

Deadly

8

1 HOUR 5 MINUTES

TIME TAKEN............................

9

1 HOUR 5 MINUTES

TIME TAKEN..............................

Deadly

10

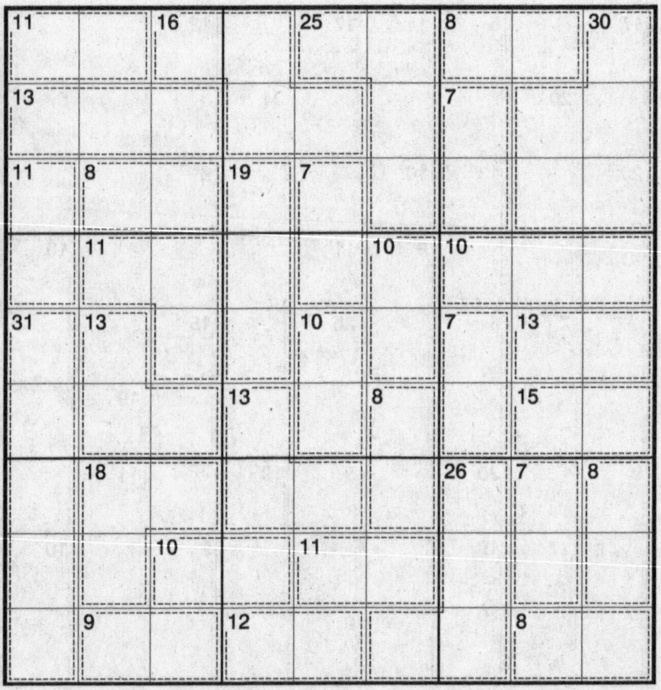

1 HOUR 5 MINUTES

TIME TAKEN..........................

11

1 HOUR 5 MINUTES

TIME TAKEN..............................

Deadly

12

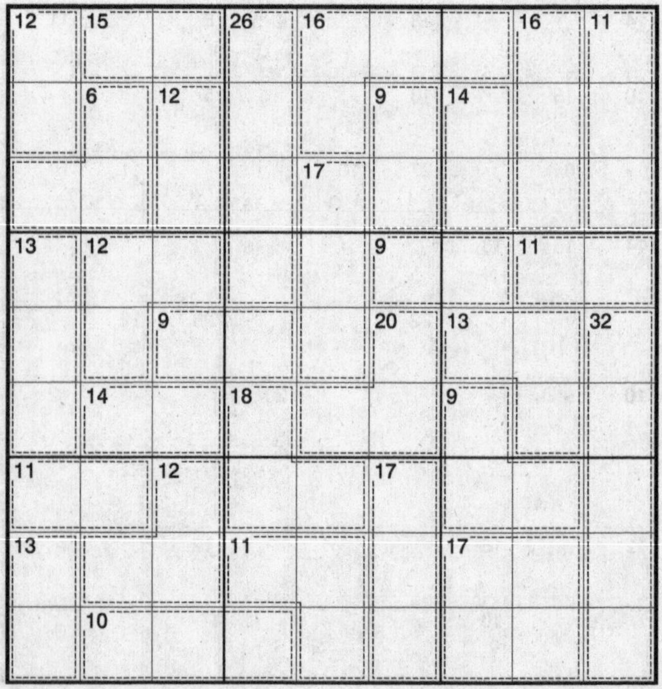

1 HOUR 5 MINUTES

TIME TAKEN..........................

13

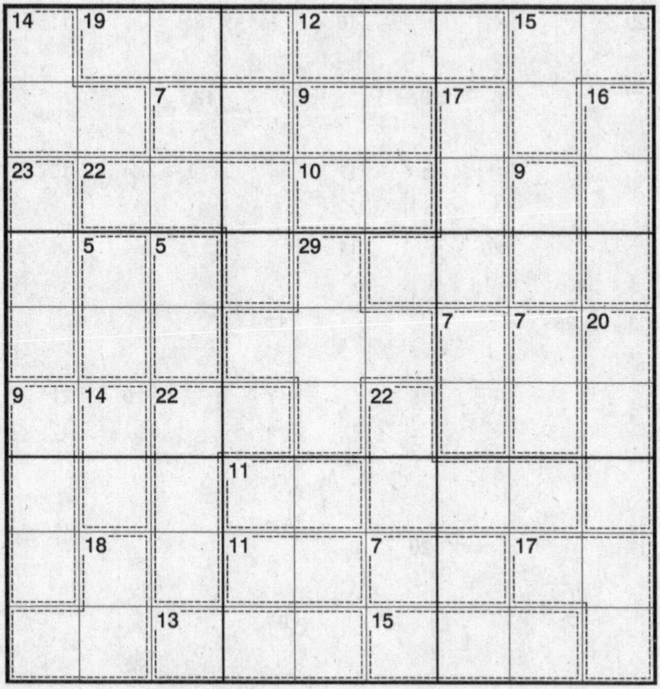

1 HOUR 5 MINUTES

TIME TAKEN..............................

Deadly

14

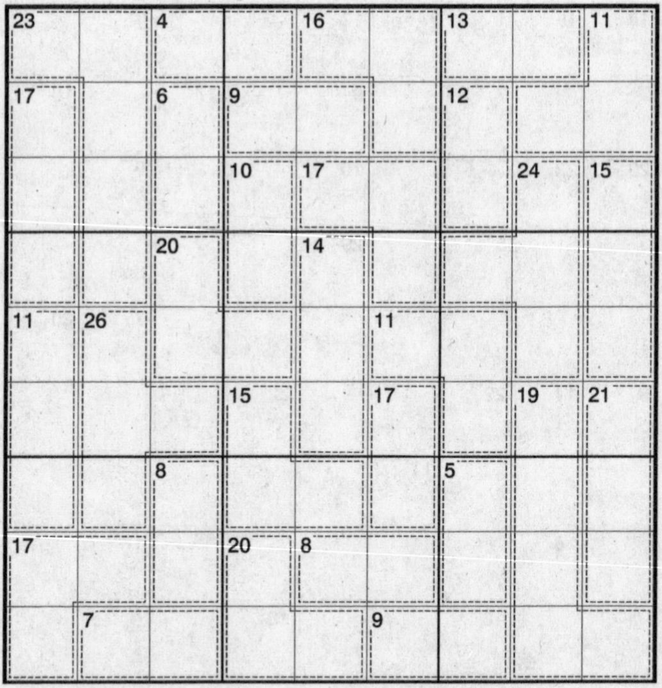

1 HOUR 5 MINUTES

TIME TAKEN..........................

15

1 HOUR 5 MINUTES

TIME TAKEN..........................

Deadly

16

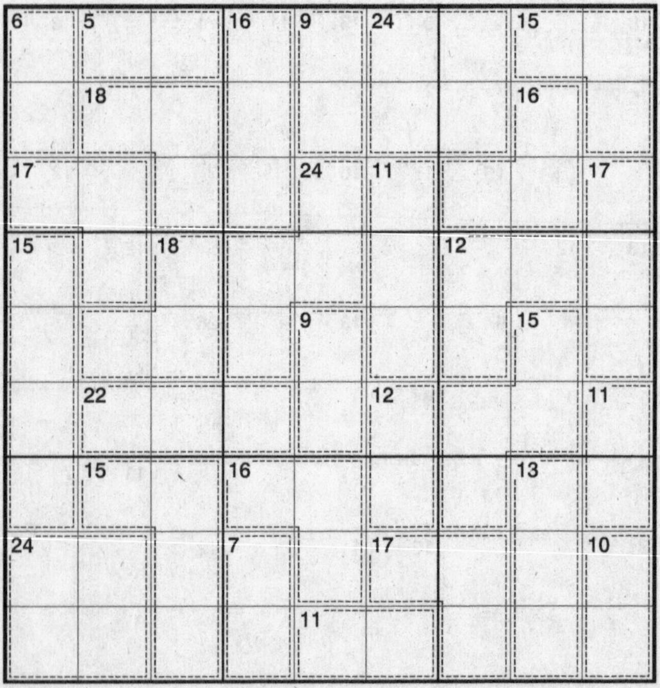

1 HOUR 5 MINUTES

TIME TAKEN..........................

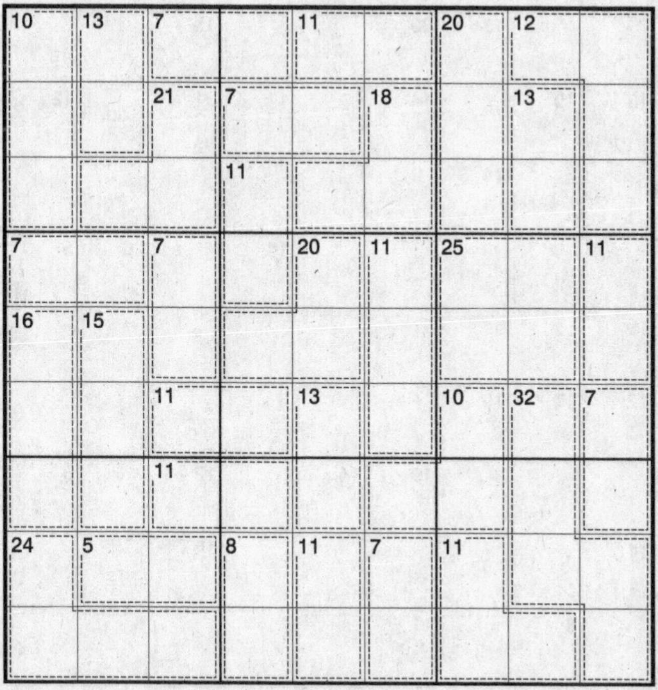

1 HOUR 5 MINUTES

TIME TAKEN.........................

Deadly

18

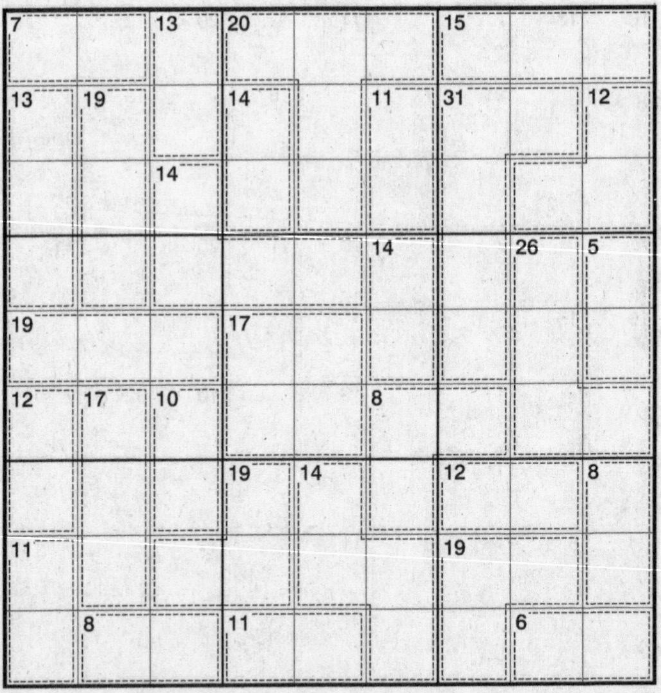

1 HOUR 5 MINUTES

TIME TAKEN..........................

19

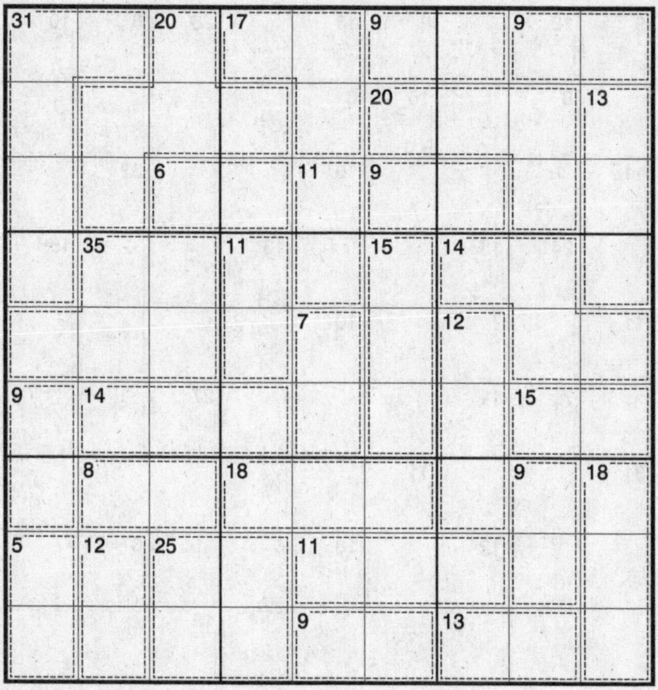

1 HOUR 5 MINUTES

TIME TAKEN.............................

Deadly

20

1 HOUR 5 MINUTES

TIME TAKEN..............................

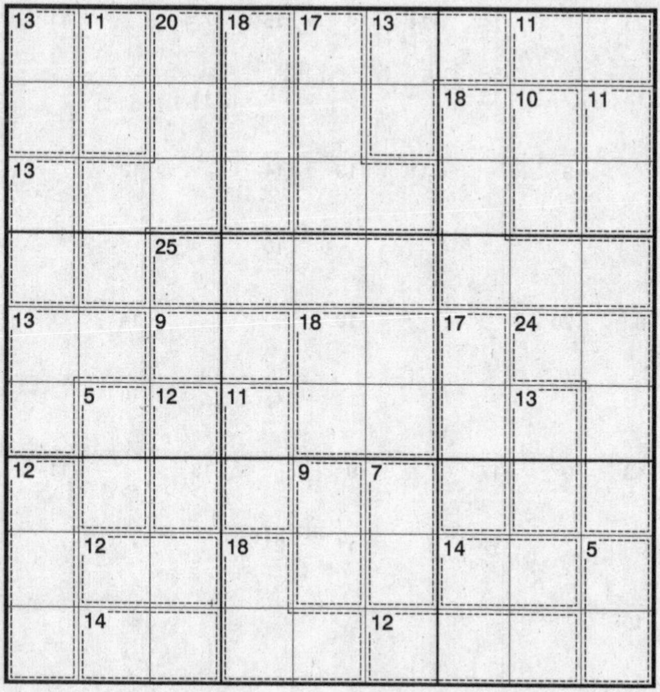

1 HOUR 5 MINUTES

TIME TAKEN.............................

Deadly

22

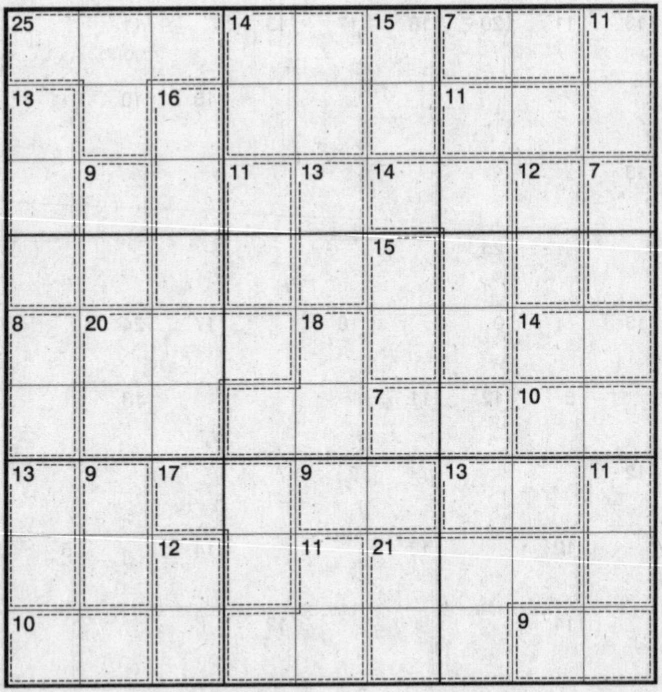

1 HOUR 5 MINUTES

TIME TAKEN.........................

23

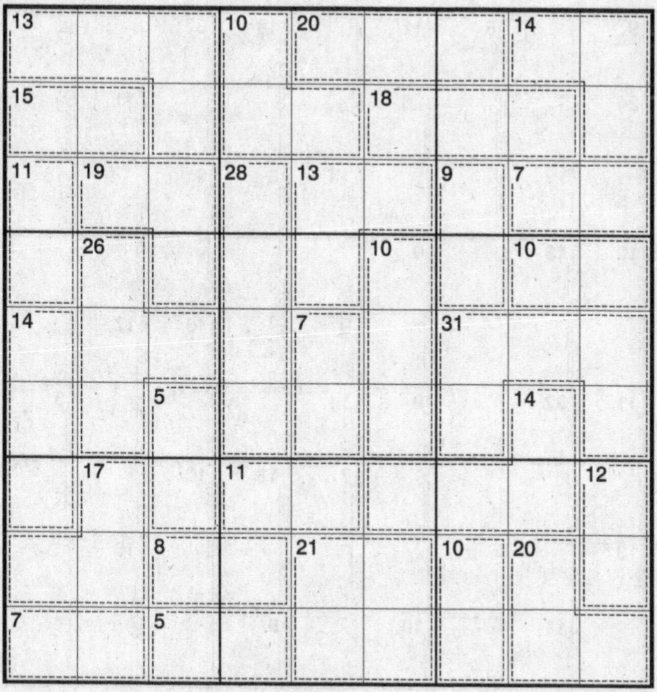

1 HOUR 5 MINUTES

TIME TAKEN............................

Deadly

24

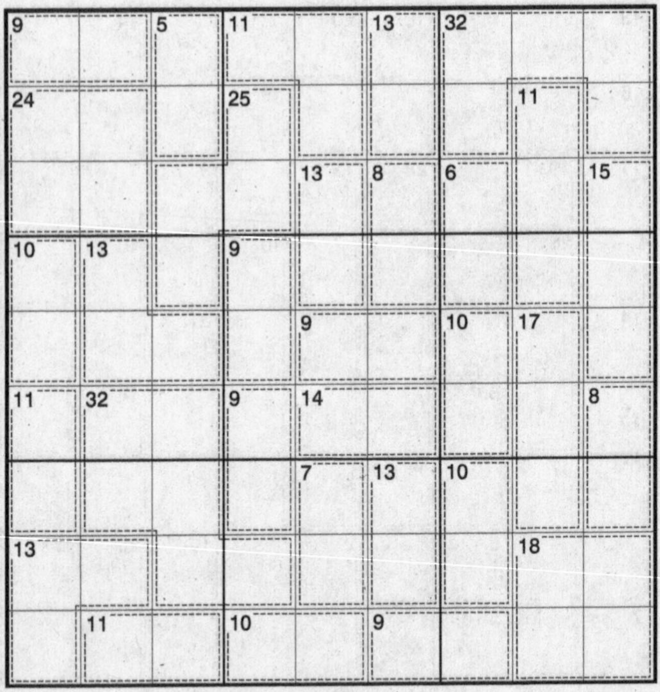

1 HOUR 5 MINUTES

TIME TAKEN............................

25

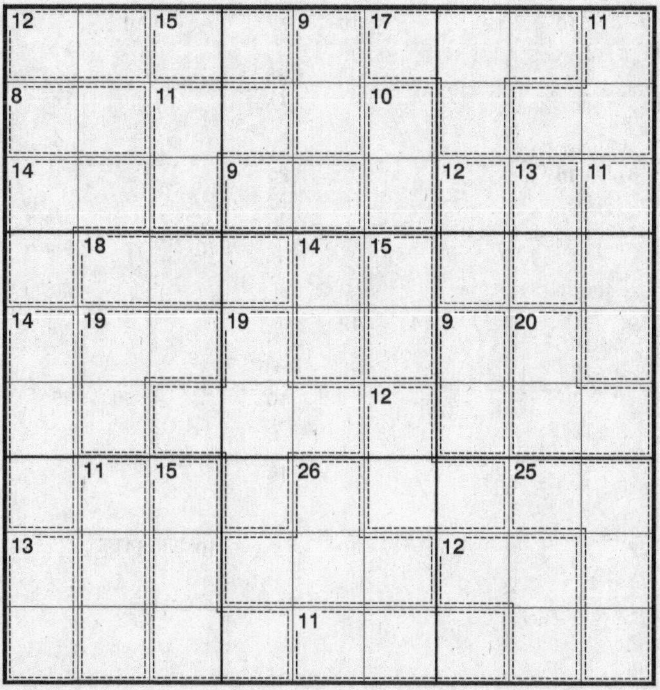

1 HOUR 5 MINUTES

TIME TAKEN..........................

Deadly

26

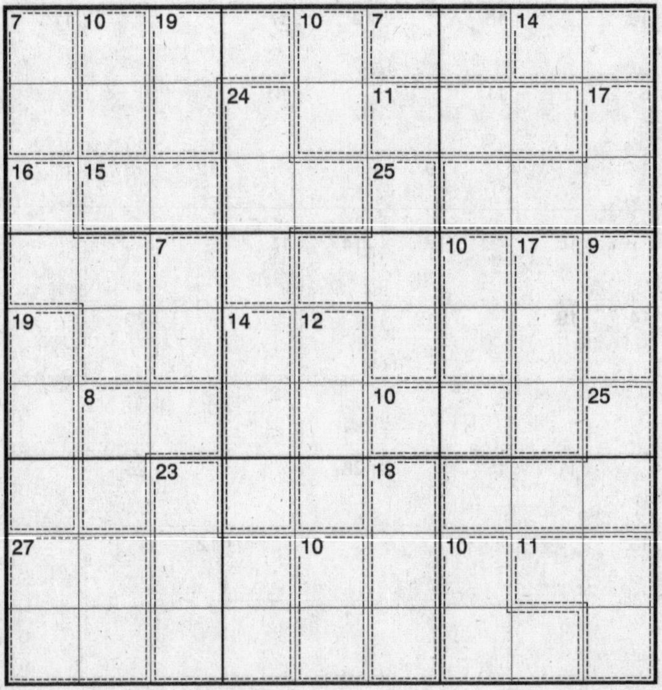

1 HOUR 5 MINUTES

TIME TAKEN............................

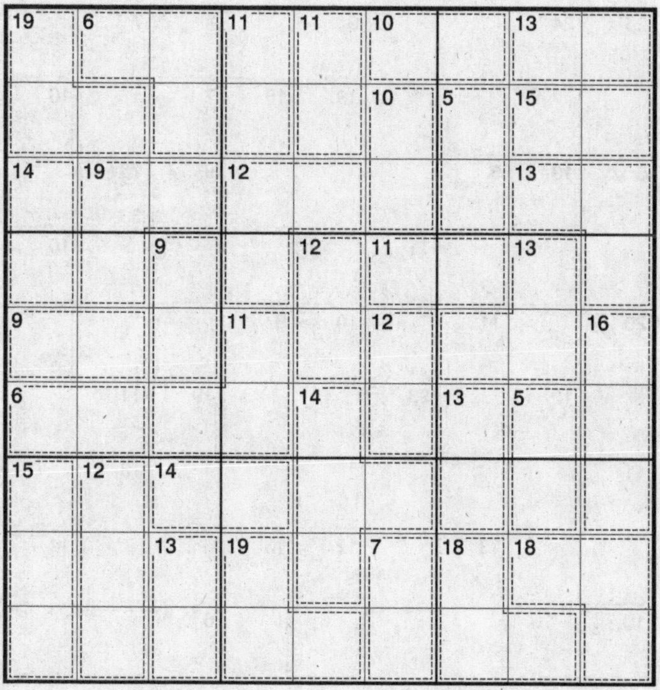

1 HOUR 5 MINUTES

TIME TAKEN.........................

Deadly

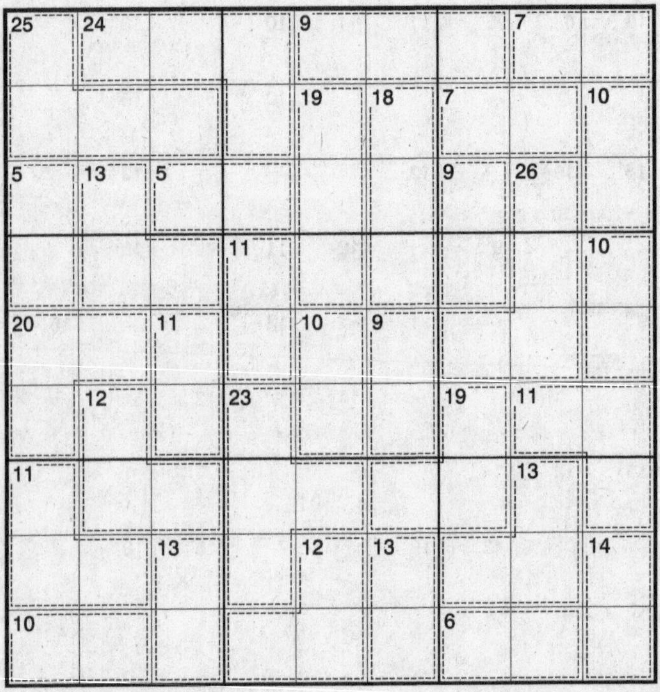

1 HOUR 5 MINUTES

TIME TAKEN..........................

29

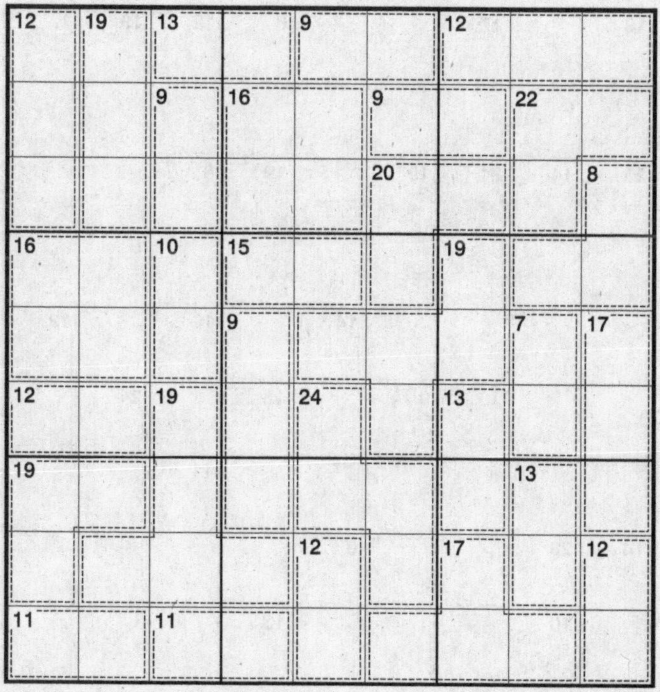

1 HOUR 5 MINUTES

TIME TAKEN..............................

Deadly

30

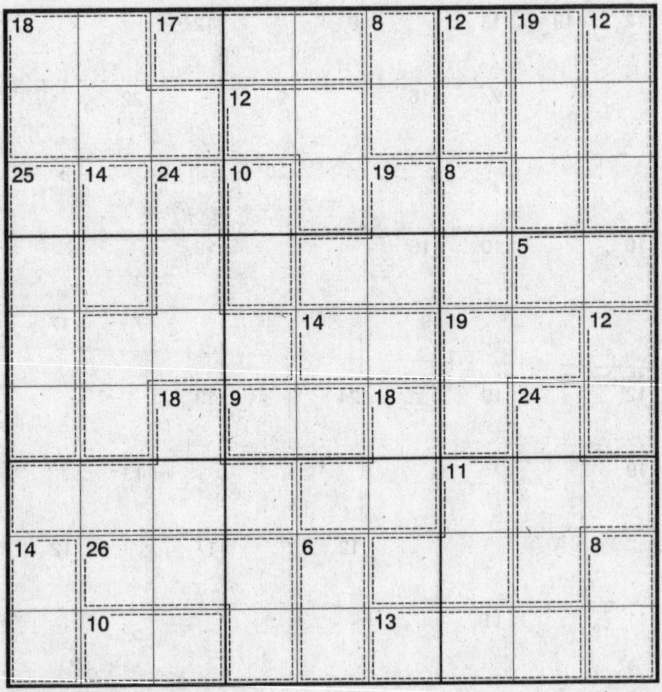

1 HOUR 5 MINUTES

TIME TAKEN.........................

31

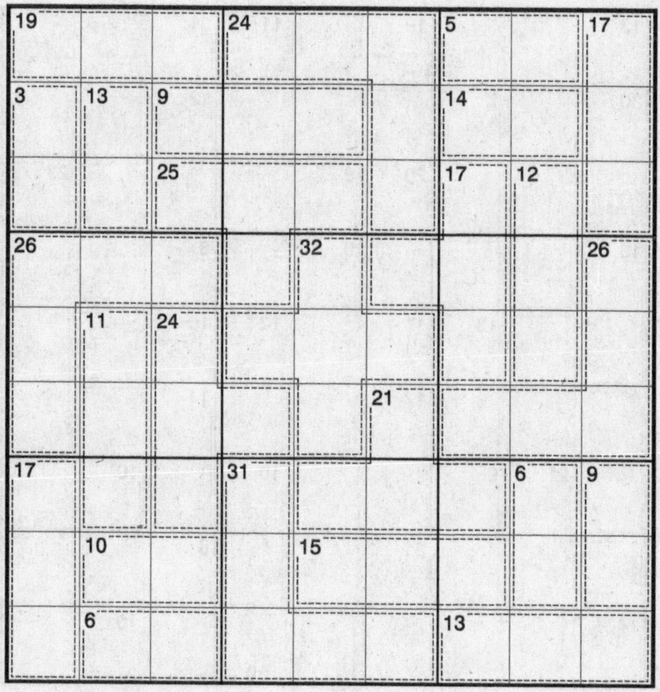

1 HOUR 10 MINUTES

TIME TAKEN..........................

Deadly

32

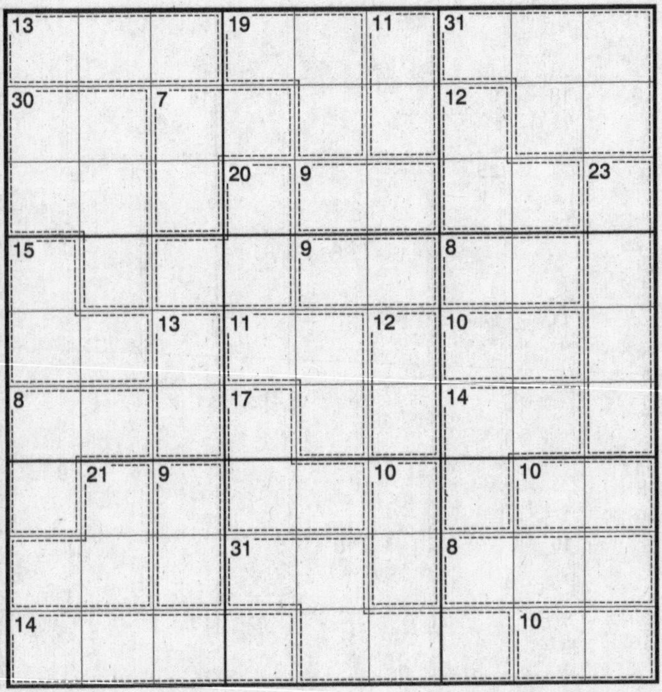

1 HOUR 10 MINUTES

TIME TAKEN........................

33

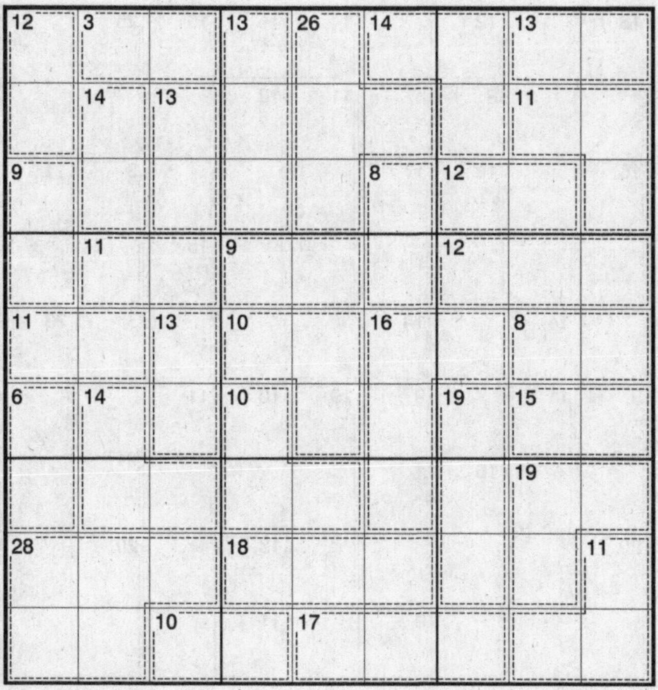

1 HOUR 10 MINUTES

TIME TAKEN.............................

Deadly

34

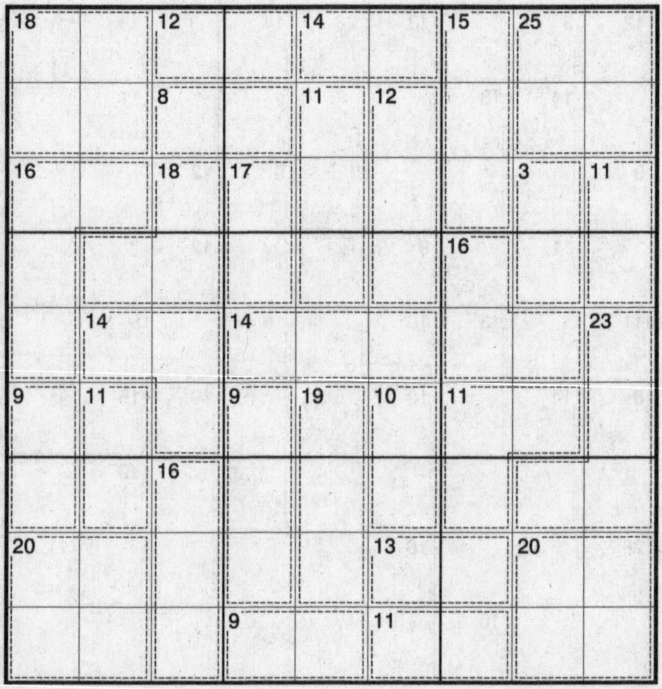

1 HOUR 10 MINUTES

TIME TAKEN..........................

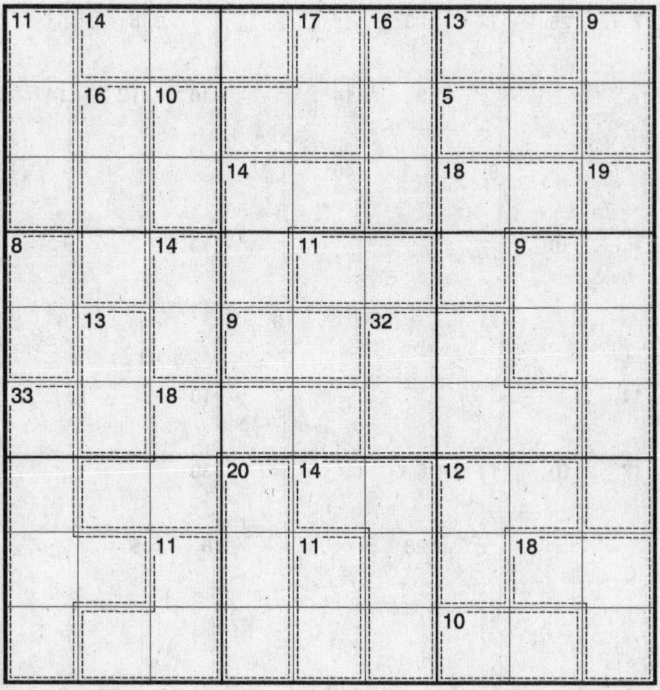

1 HOUR 10 MINUTES

TIME TAKEN............................

Deadly

36

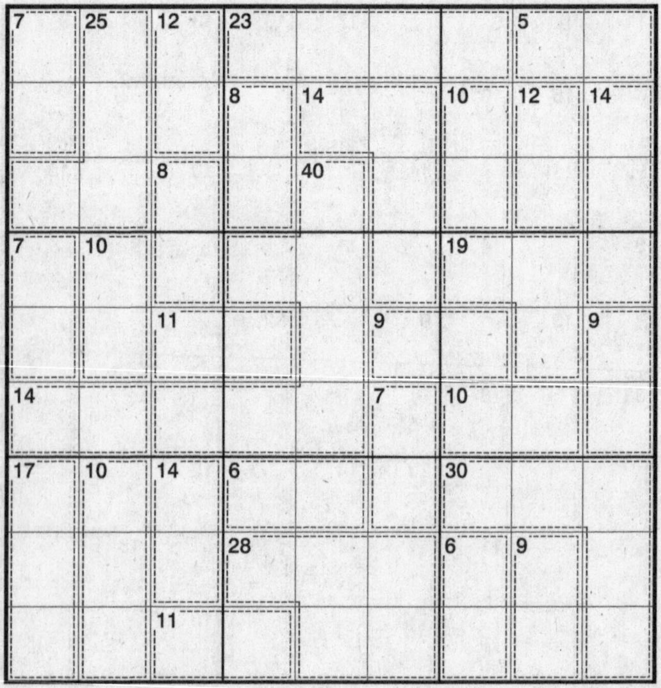

1 HOUR 10 MINUTES

TIME TAKEN.........................

37

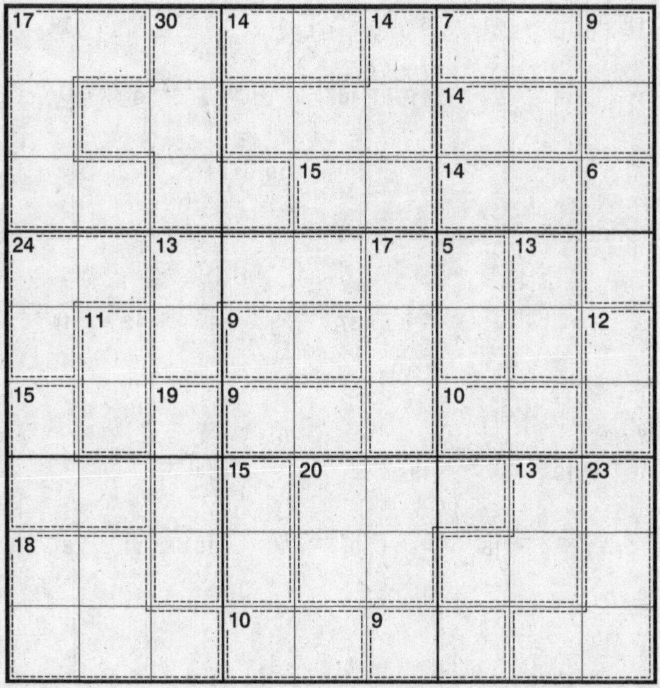

1 HOUR 10 MINUTES

TIME TAKEN............................

Deadly

38

1 HOUR 10 MINUTES

TIME TAKEN..........................

1 HOUR 10 MINUTES

TIME TAKEN.............................

Deadly

40

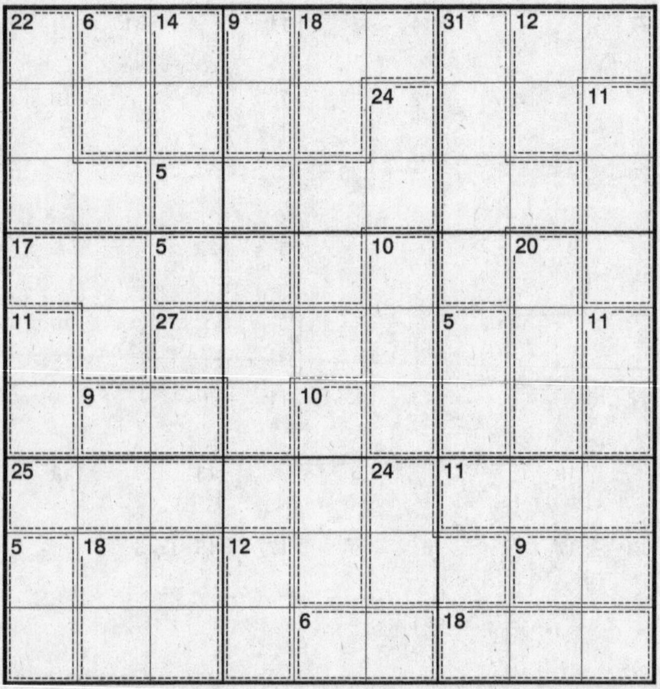

1 HOUR 10 MINUTES

TIME TAKEN..........................

41

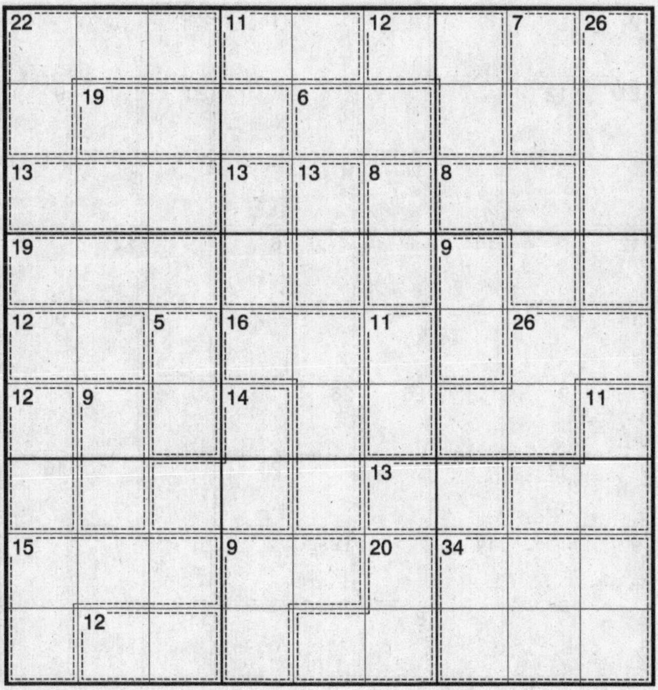

1 HOUR 10 MINUTES

TIME TAKEN..........................

Deadly

42

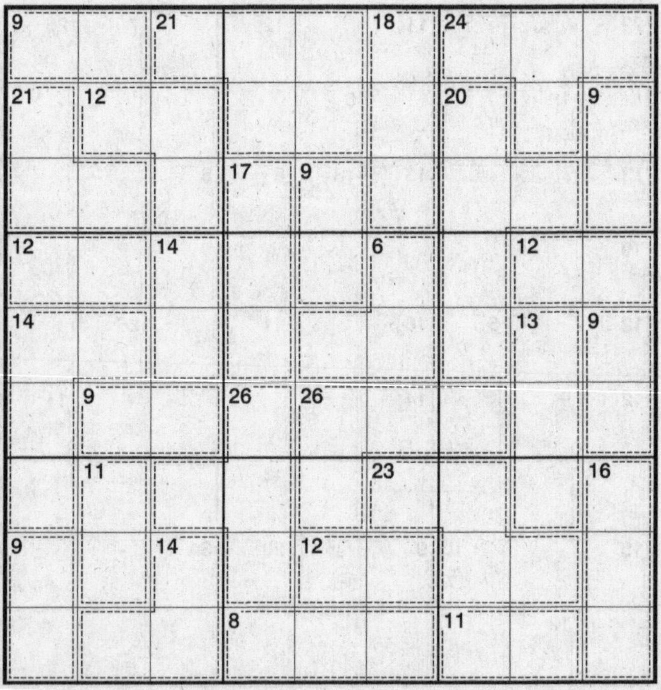

1 HOUR 10 MINUTES

TIME TAKEN..........................

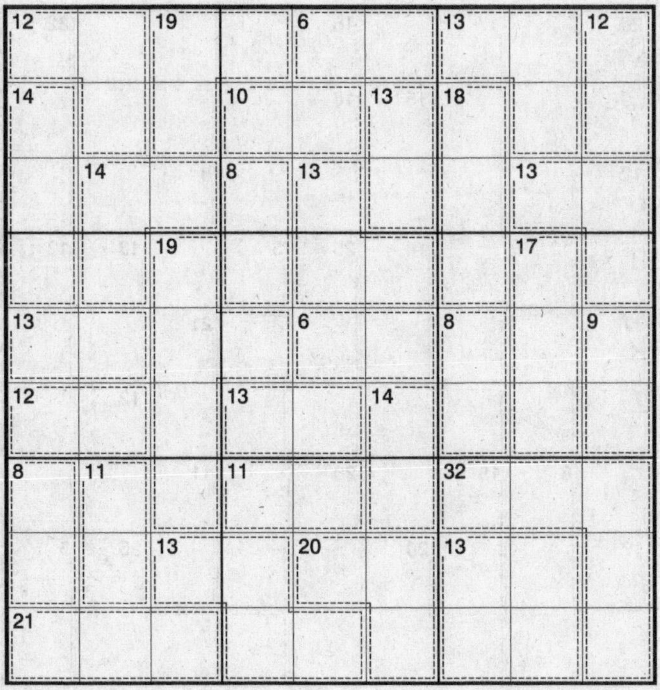

1 HOUR 10 MINUTES

TIME TAKEN............................

44

1 HOUR 10 MINUTES

TIME TAKEN..........................

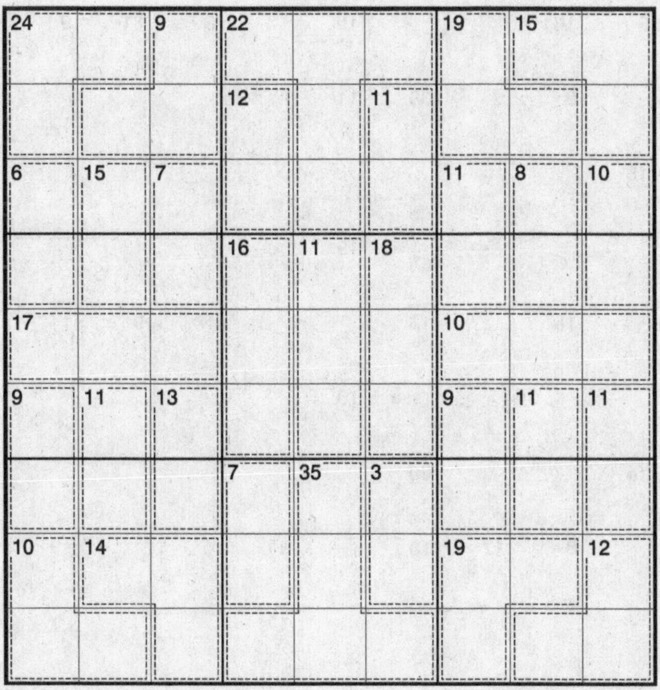

1 HOUR 10 MINUTES

TIME TAKEN..........................

46

1 HOUR 10 MINUTES

TIME TAKEN............................

47

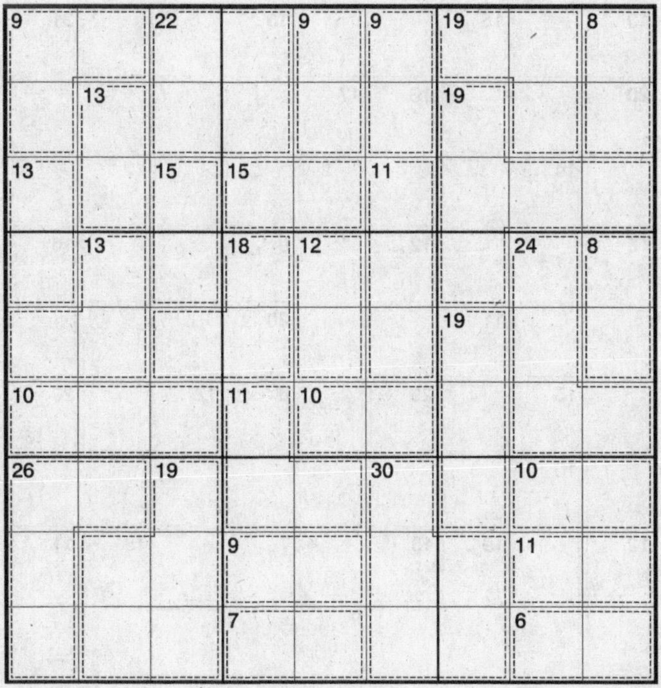

1 HOUR 10 MINUTES

TIME TAKEN..........................

Deadly

48

1 HOUR 10 MINUTES

TIME TAKEN..........................

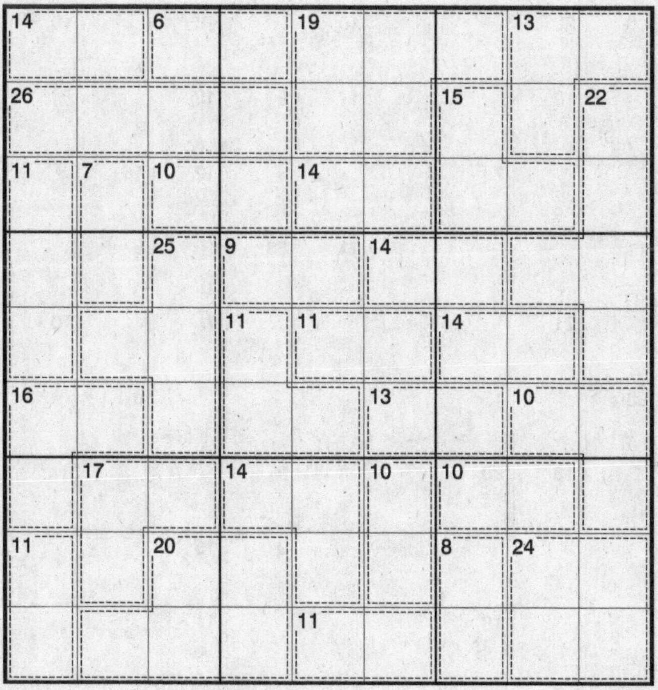

1 HOUR 10 MINUTES

TIME TAKEN..............................

Deadly

50

1 HOUR 10 MINUTES

TIME TAKEN.........................

1 HOUR 10 MINUTES

TIME TAKEN..........................

Deadly

52

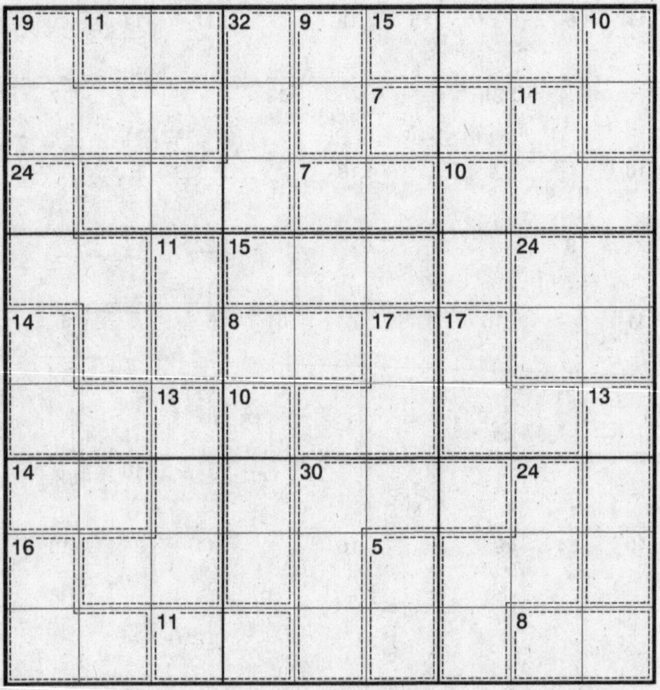

1 HOUR 10 MINUTES

TIME TAKEN..........................

53

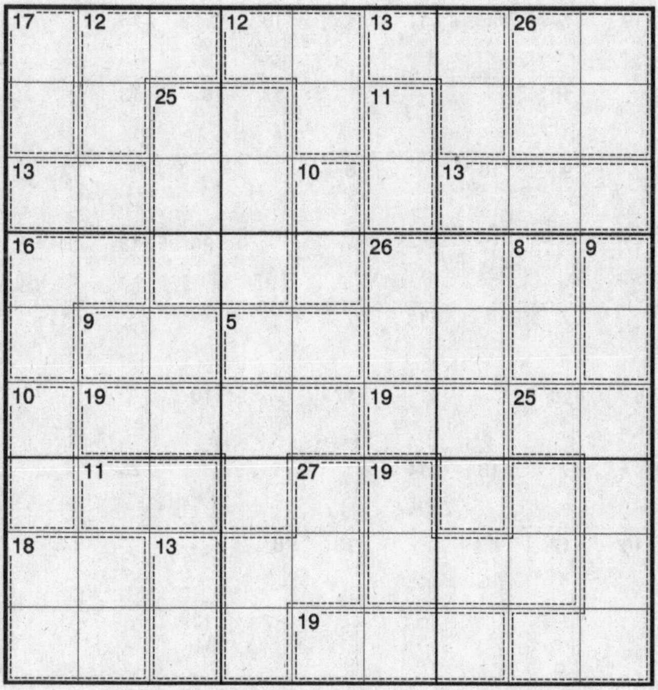

1 HOUR 10 MINUTES

TIME TAKEN..........................

Deadly

54

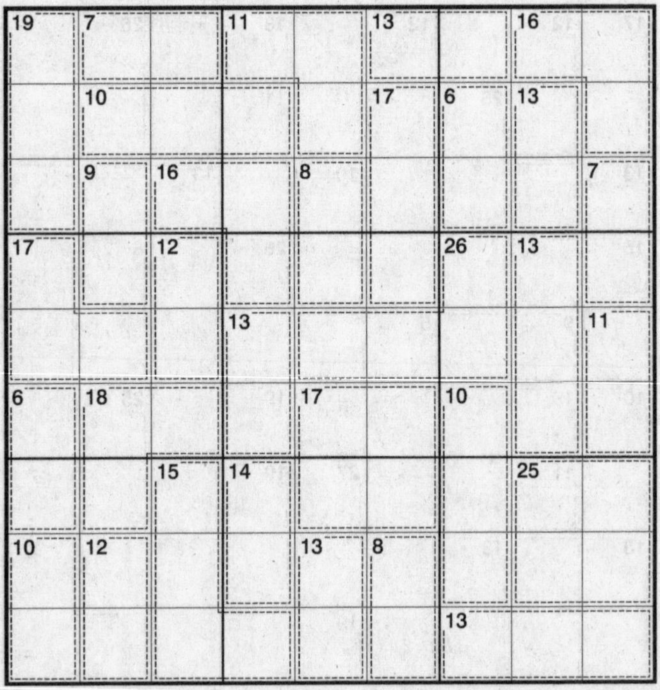

1 HOUR 10 MINUTES

TIME TAKEN..........................

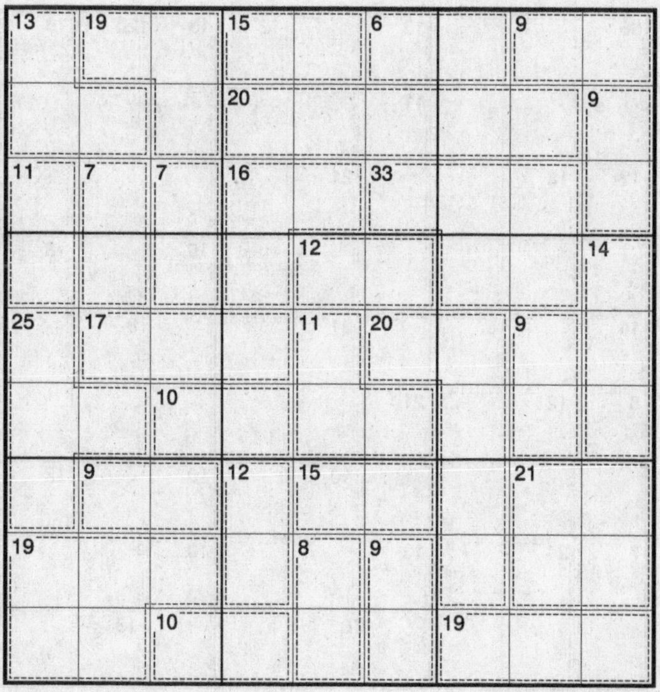

1 HOUR 10 MINUTES

TIME TAKEN..........................

56

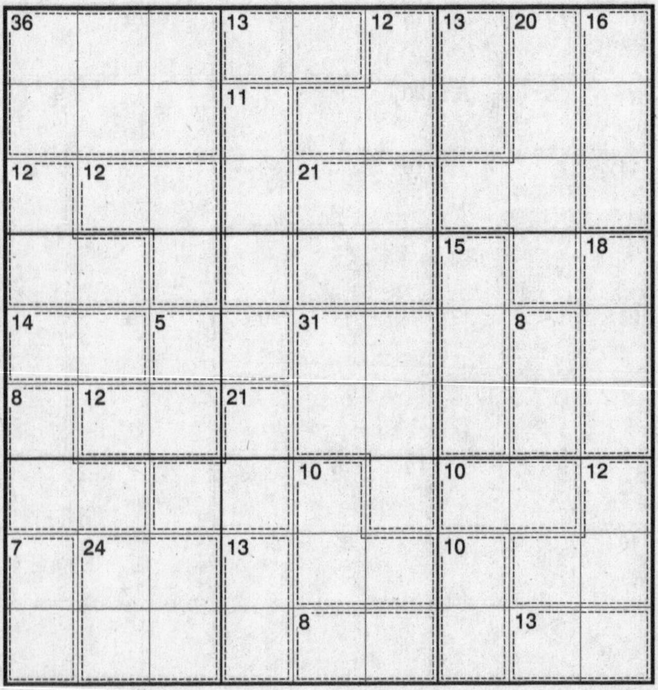

1 HOUR 15 MINUTES

TIME TAKEN..........................

57

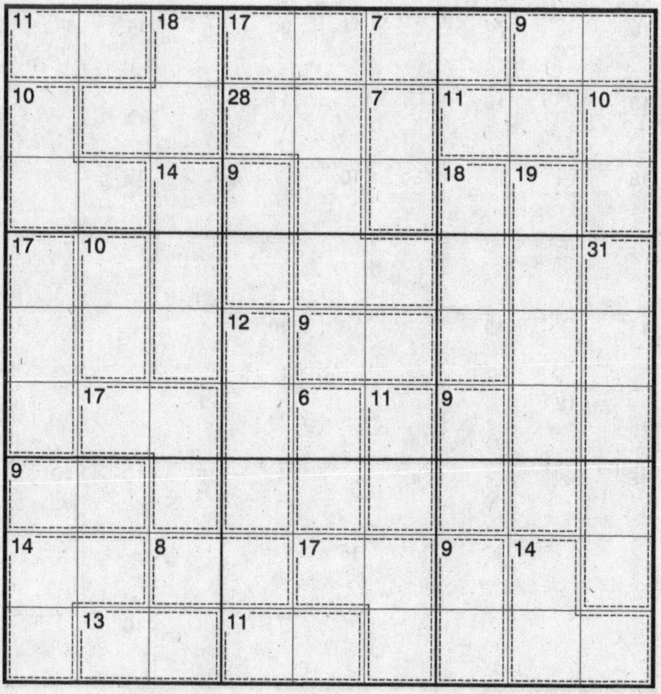

1 HOUR 15 MINUTES

TIME TAKEN.............................

Deadly

58

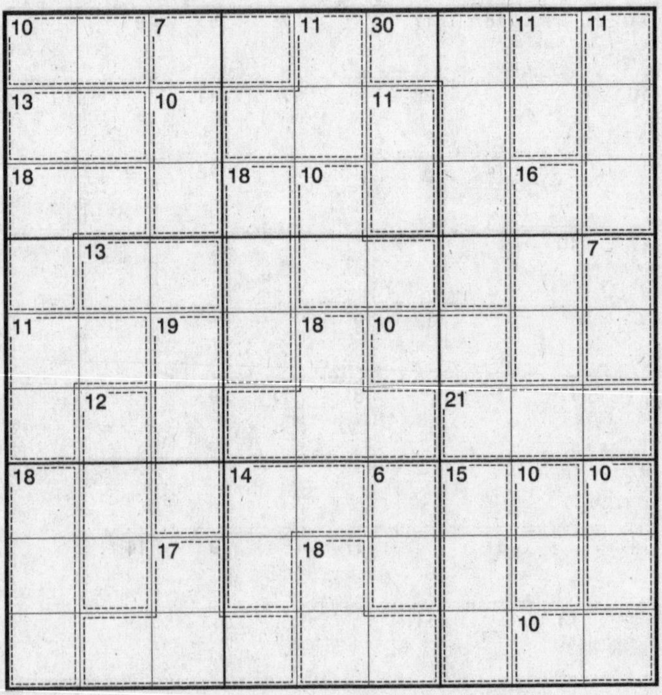

1 HOUR 15 MINUTES

TIME TAKEN..........................

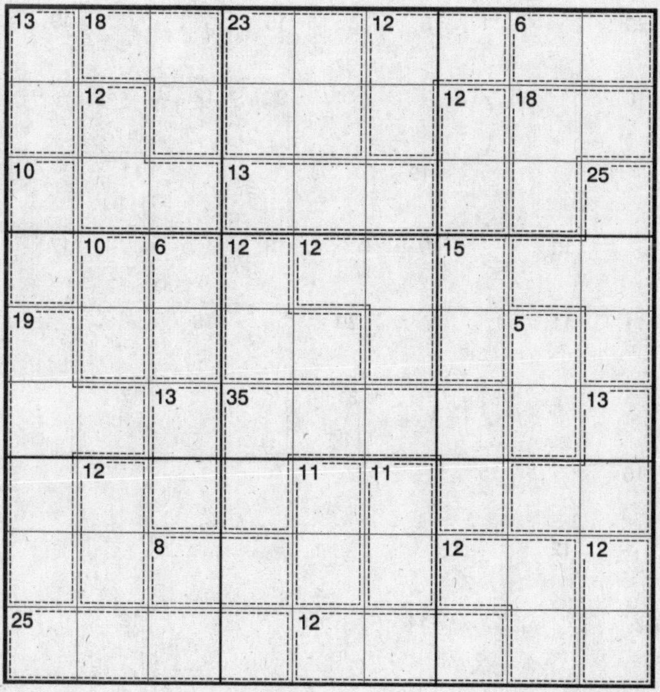

1 HOUR 15 MINUTES

TIME TAKEN..........................

60

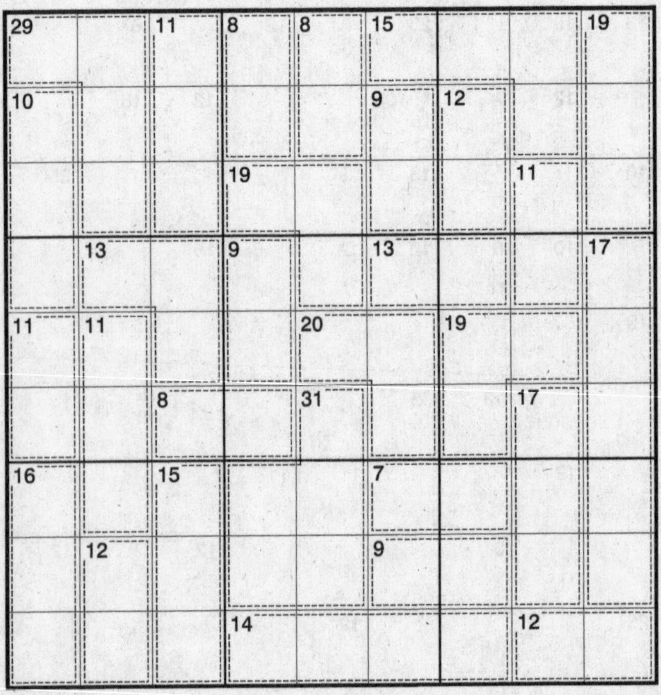

1 HOUR 15 MINUTES

TIME TAKEN..........................

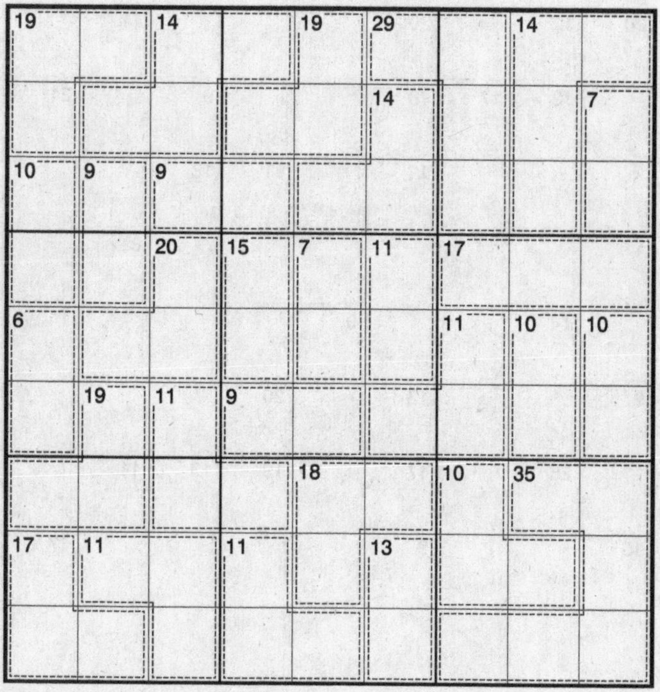

1 HOUR 15 MINUTES

TIME TAKEN.........................

Deadly

62

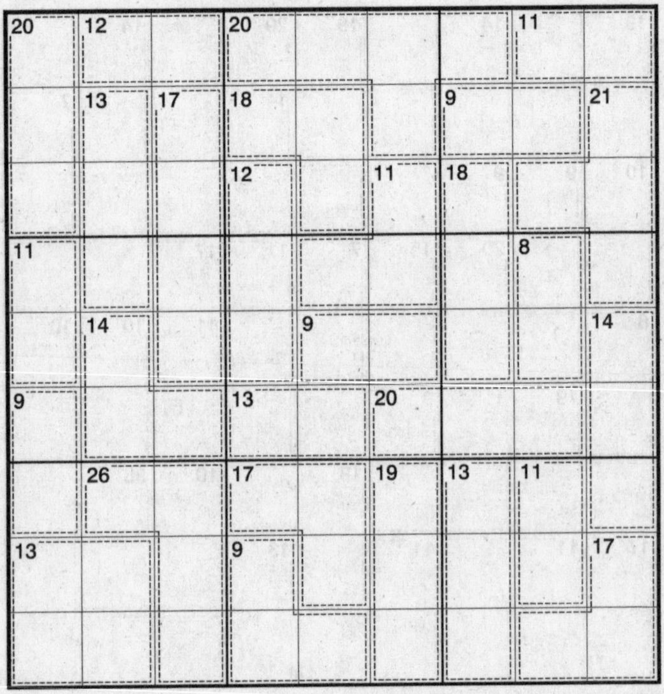

1 HOUR 15 MINUTES

TIME TAKEN..........................

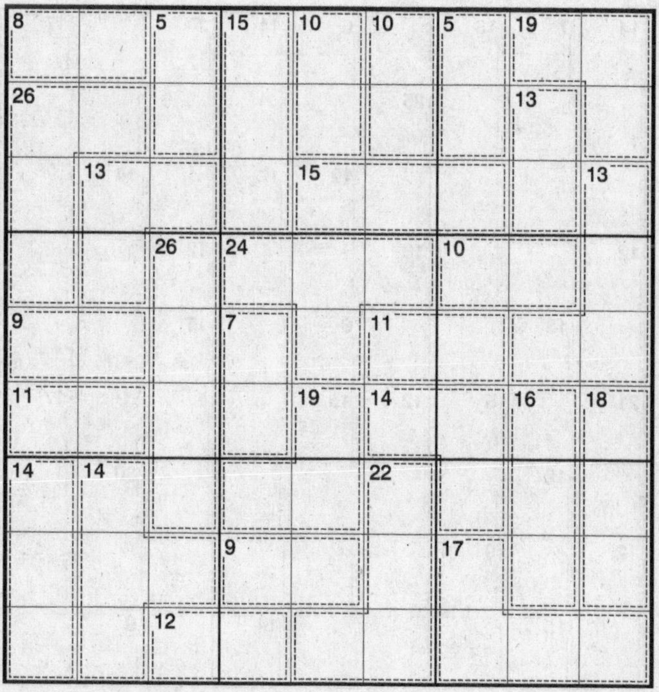

1 HOUR 15 MINUTES

TIME TAKEN.........................

Deadly

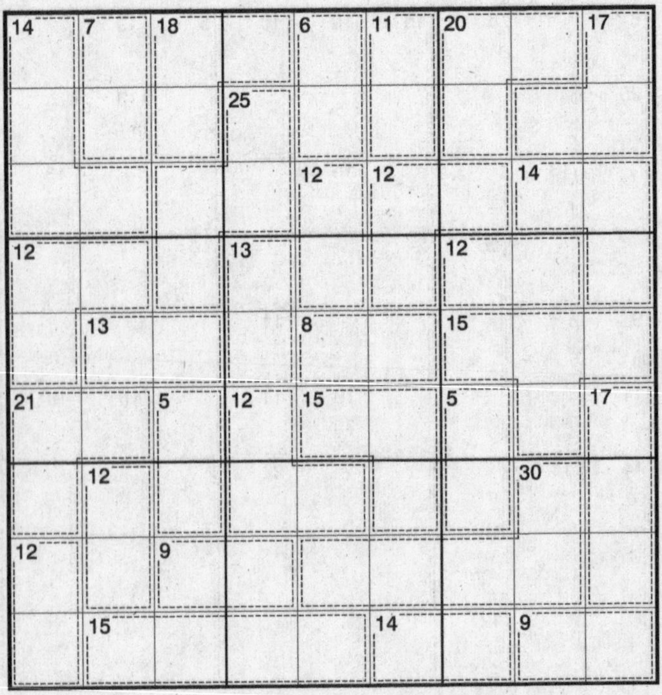

1 HOUR 15 MINUTES

TIME TAKEN..........................

65

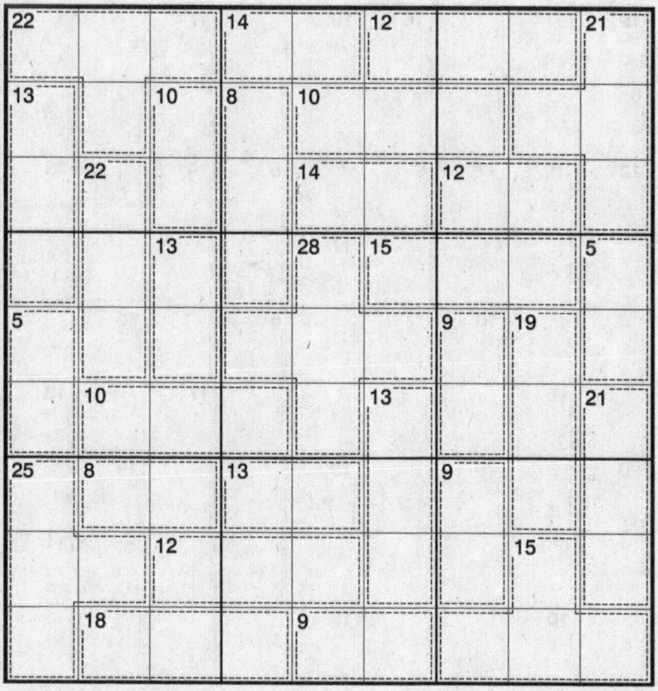

1 HOUR 15 MINUTES

TIME TAKEN..........................

Deadly

66

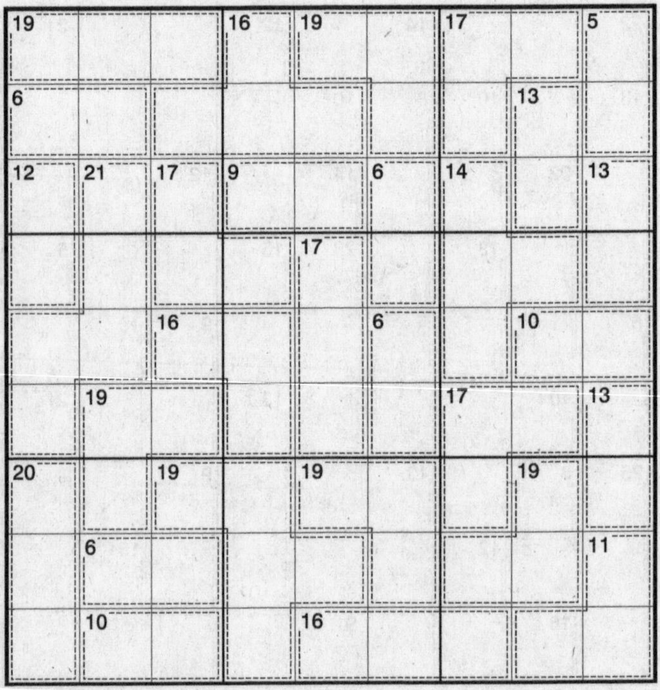

1 HOUR 15 MINUTES

TIME TAKEN.........................

67

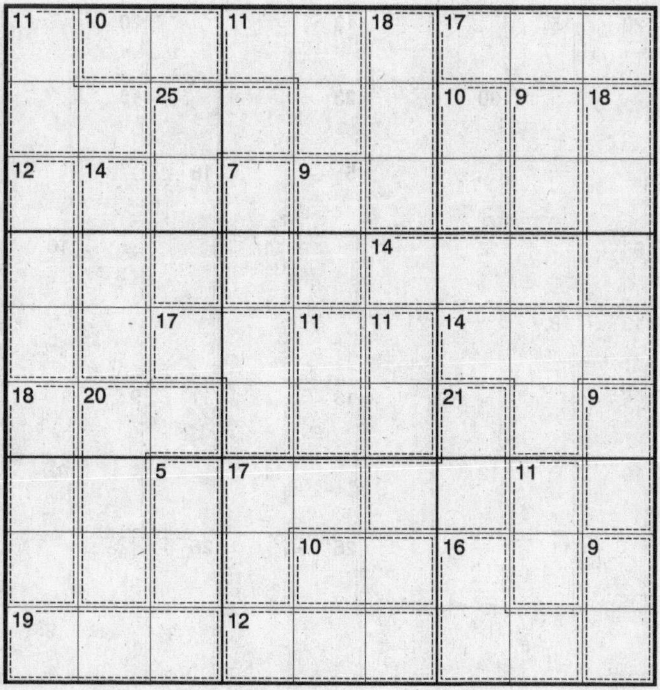

1 HOUR 15 MINUTES

TIME TAKEN..........................

Deadly

68

1 HOUR 15 MINUTES

TIME TAKEN..........................

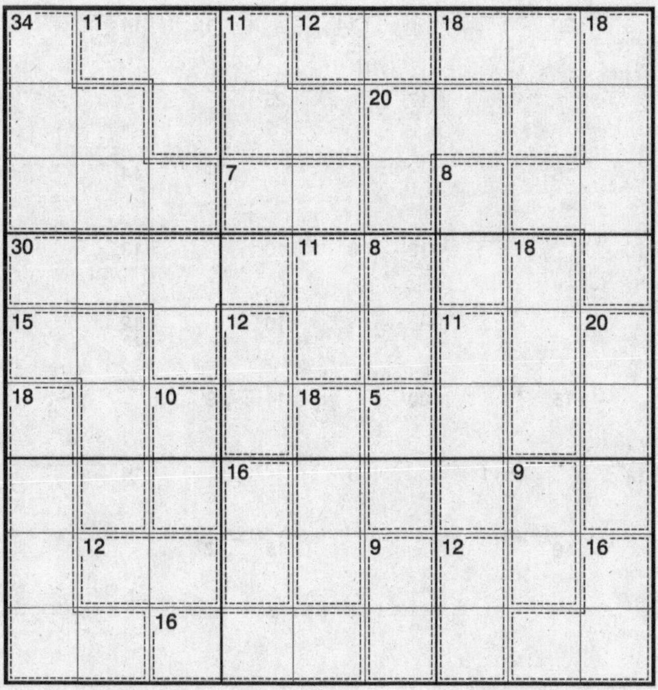

1 HOUR 15 MINUTES

TIME TAKEN..........................

70

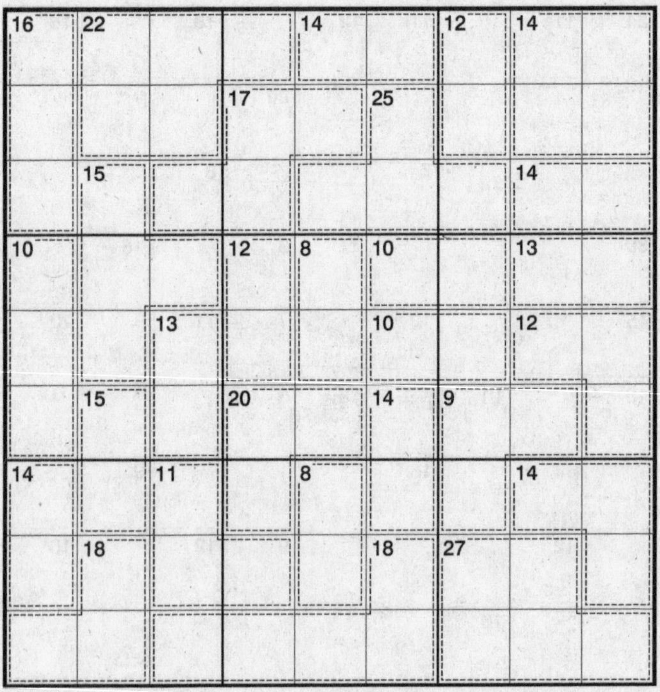

1 HOUR 15 MINUTES

TIME TAKEN..........................

71

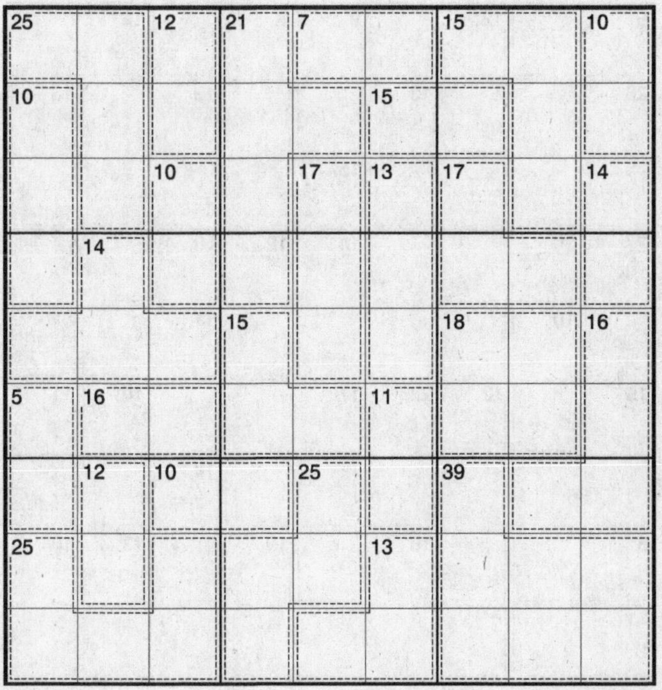

1 HOUR 15 MINUTES

TIME TAKEN............................

Deadly

72

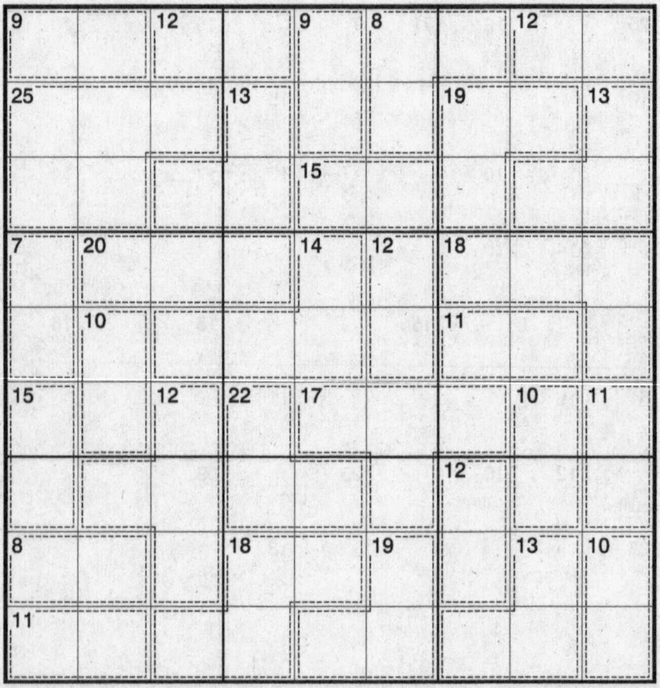

1 HOUR 15 MINUTES

TIME TAKEN...........................

73

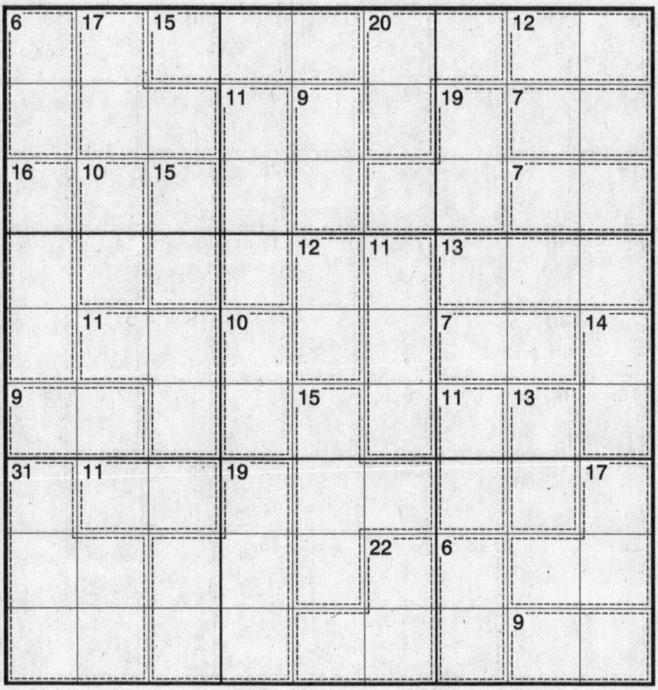

1 HOUR 15 MINUTES

TIME TAKEN..............................

Deadly

74

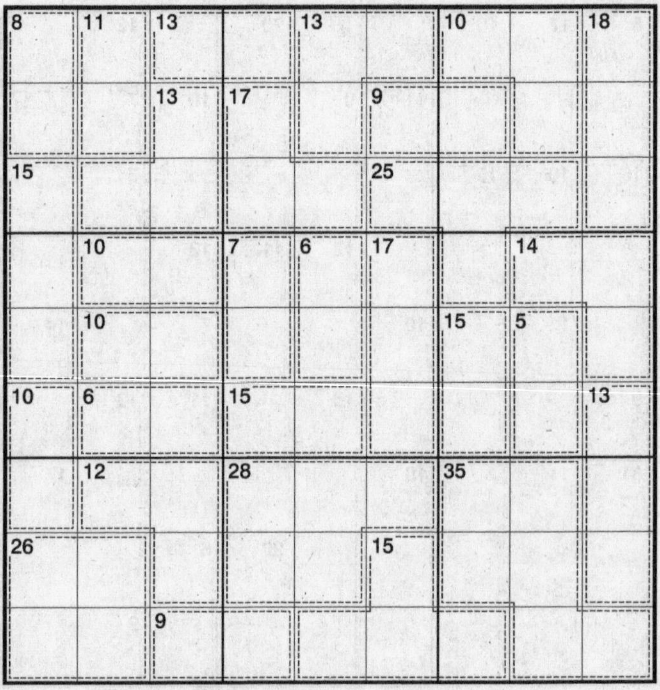

1 HOUR 15 MINUTES

TIME TAKEN..........................

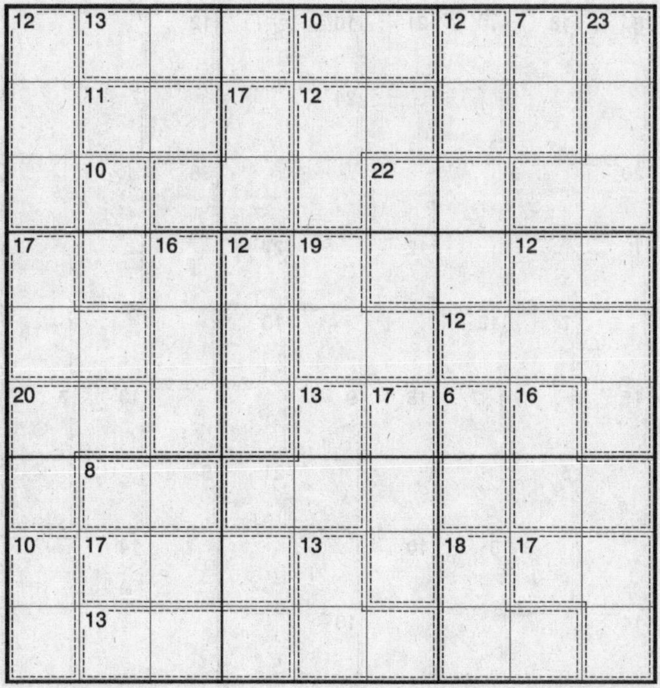

1 HOUR 15 MINUTES

TIME TAKEN...........................

76

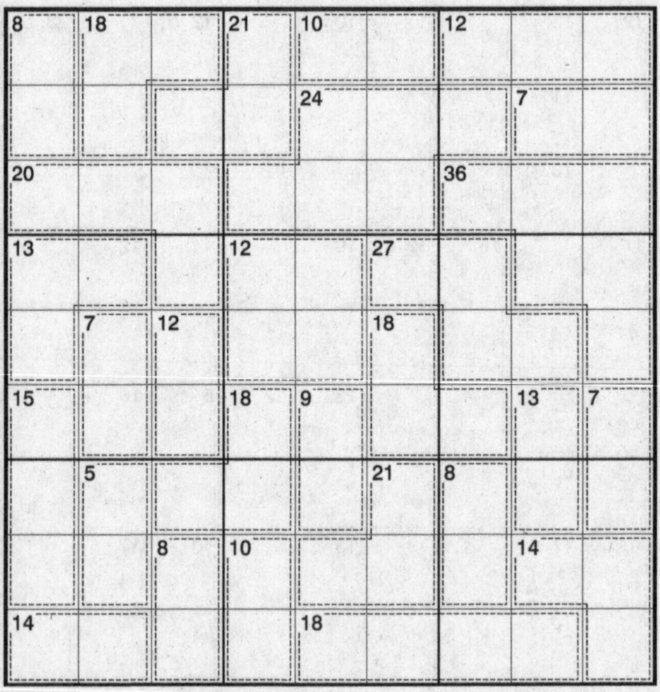

1 HOUR 15 MINUTES

TIME TAKEN..........................

77

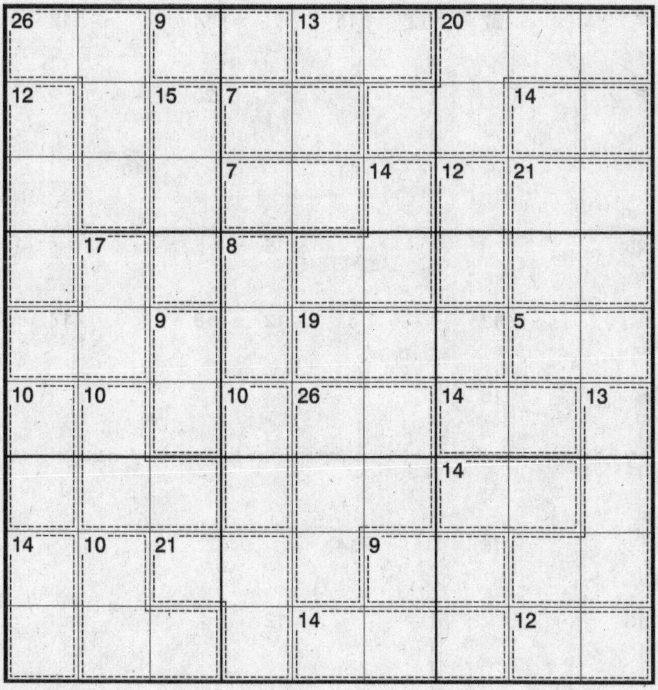

1 HOUR 20 MINUTES

TIME TAKEN.........................

Deadly

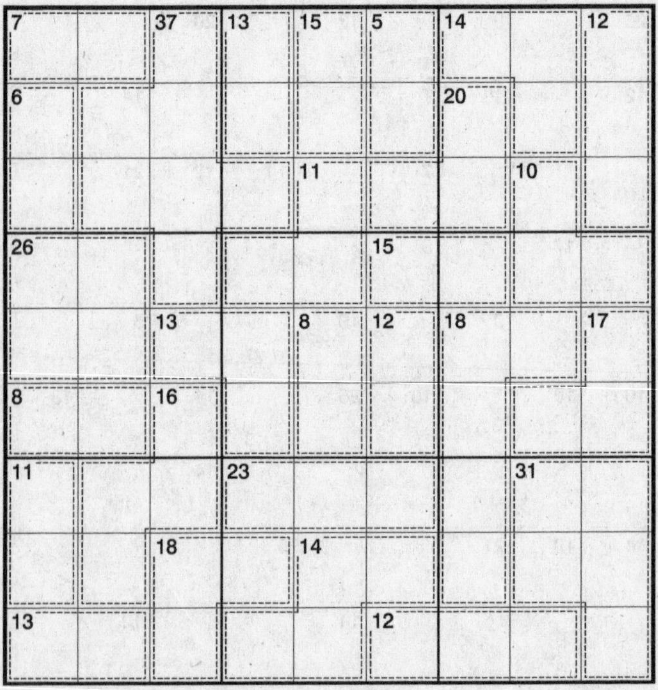

1 HOUR 20 MINUTES

TIME TAKEN.............................

79

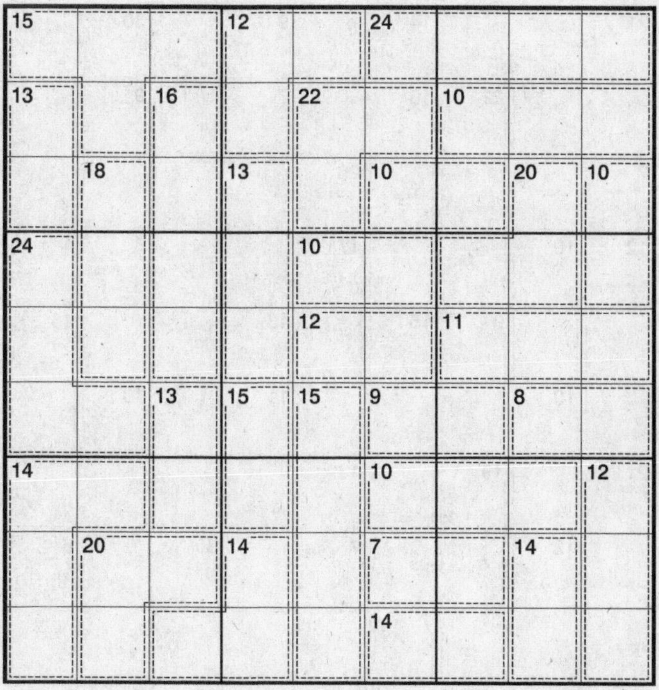

1 HOUR 20 MINUTES

TIME TAKEN..........................

Deadly

80

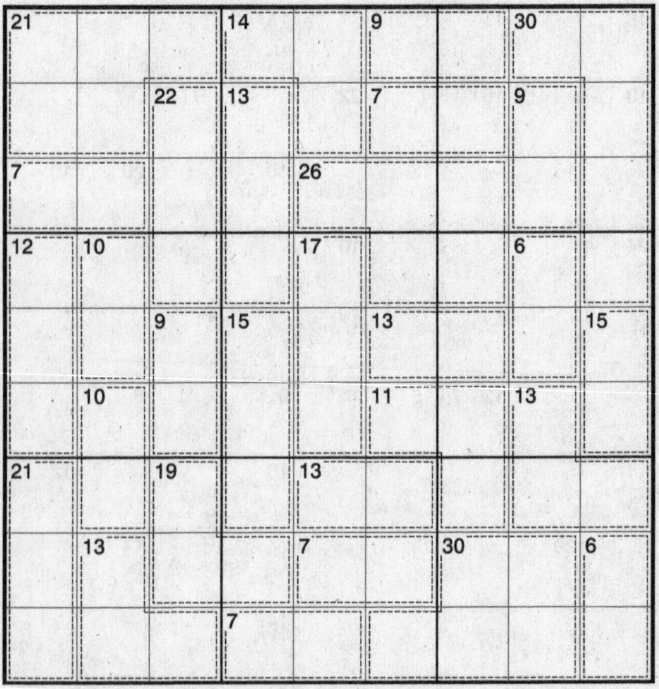

1 HOUR 20 MINUTES

TIME TAKEN..........................

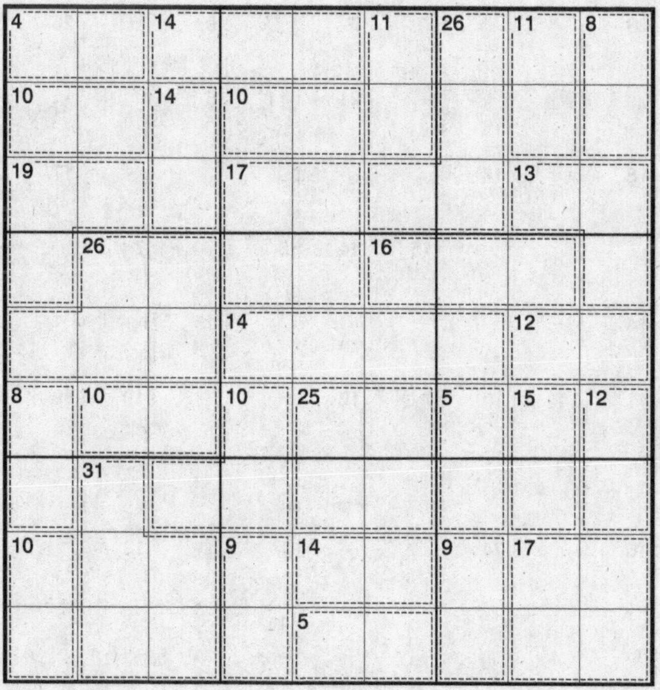

1 HOUR 20 MINUTES

TIME TAKEN.............................

Deadly

82

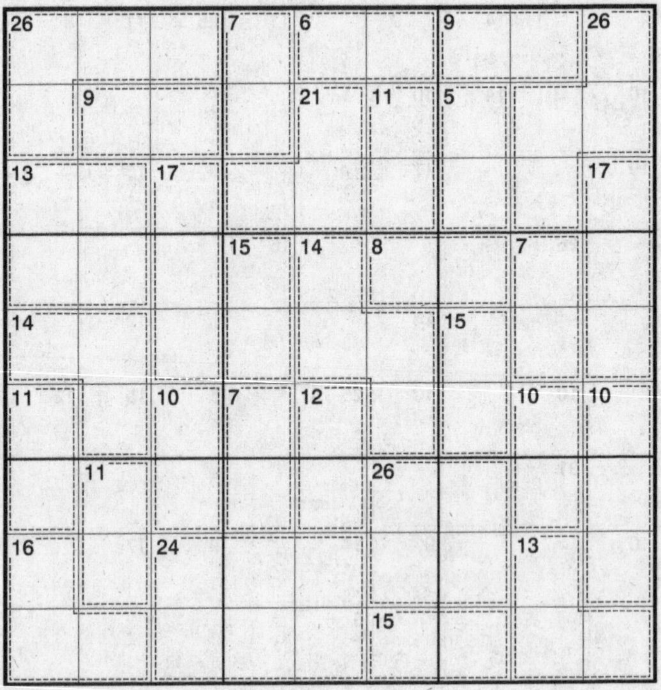

1 HOUR 20 MINUTES

TIME TAKEN..........................

83

1 HOUR 20 MINUTES

TIME TAKEN..........................

Deadly

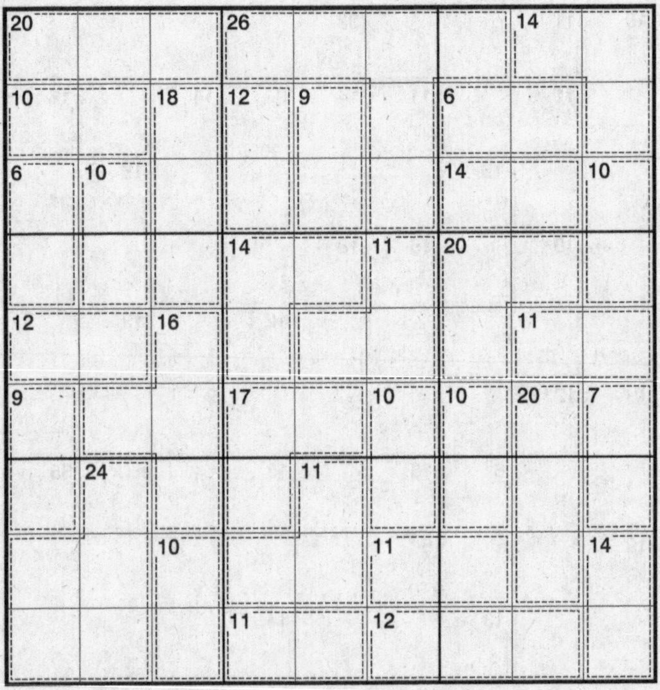

1 HOUR 20 MINUTES

TIME TAKEN..........................

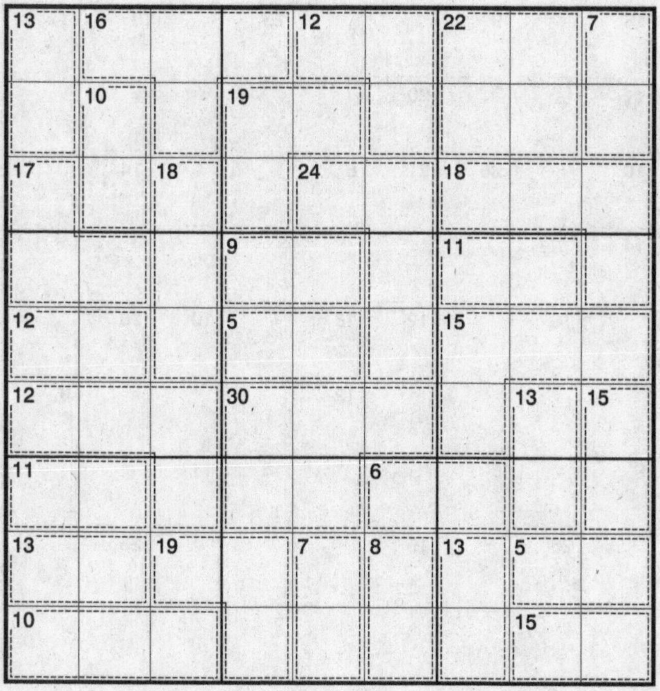

1 HOUR 20 MINUTES

TIME TAKEN..........................

Deadly

86

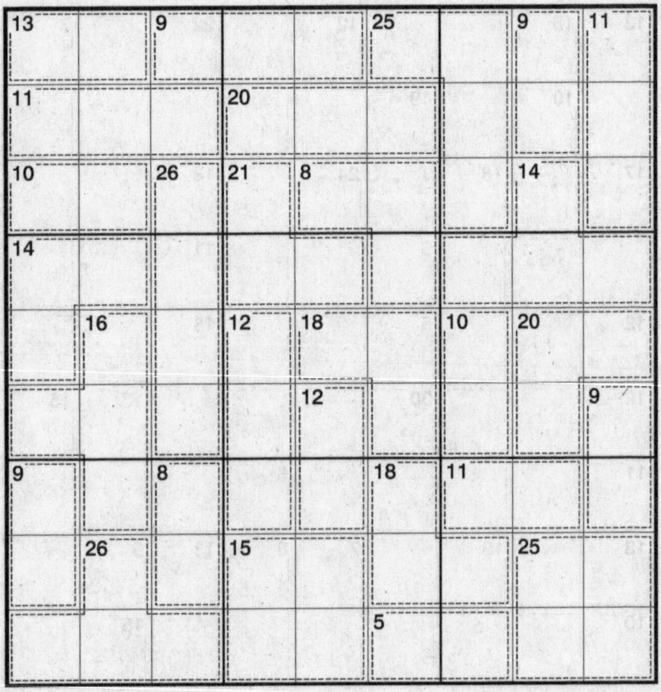

1 HOUR 20 MINUTES

TIME TAKEN..........................

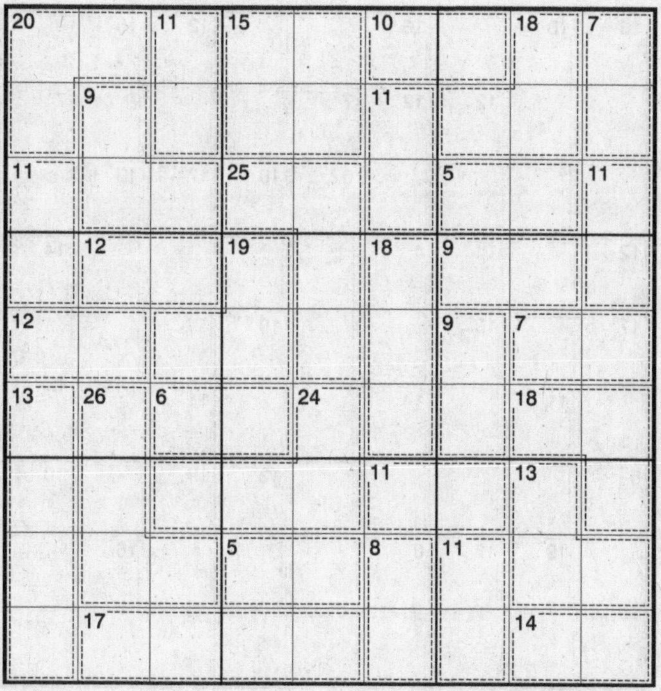

1 HOUR 20 MINUTES

TIME TAKEN..............................

Deadly

88

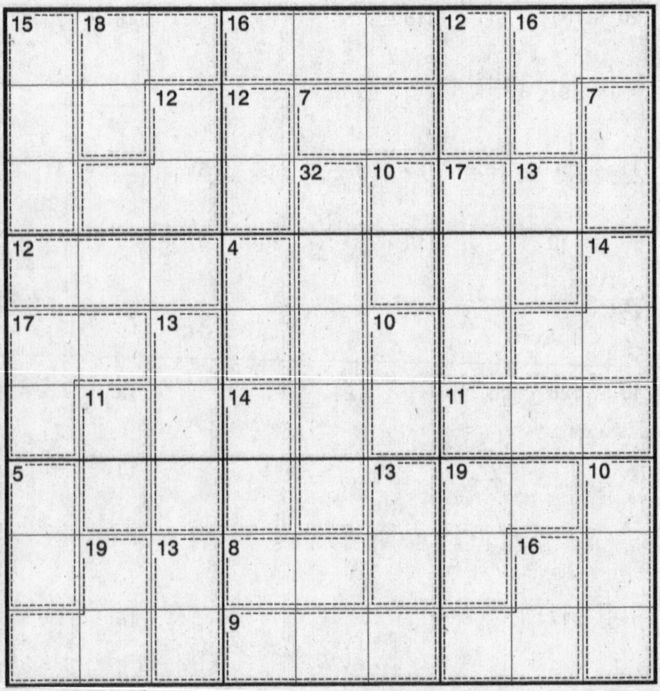

1 HOUR 20 MINUTES

TIME TAKEN..........................

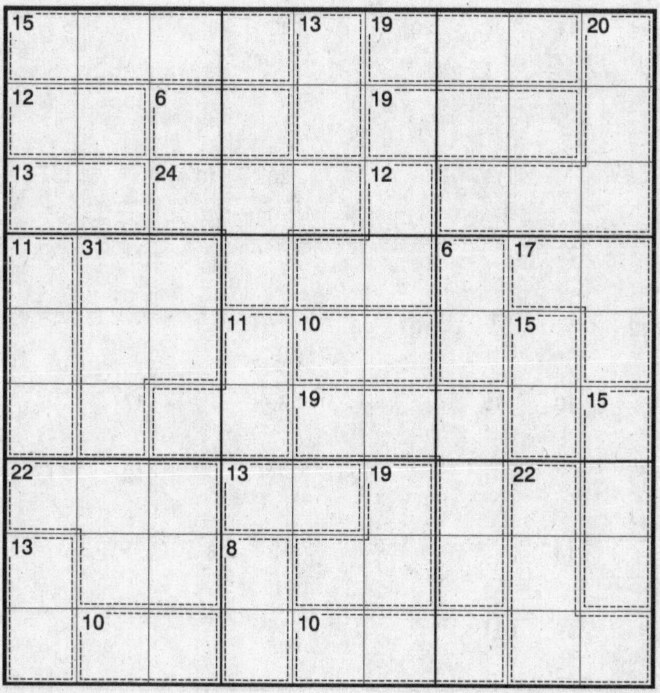

1 HOUR 20 MINUTES

TIME TAKEN..........................

Deadly

90

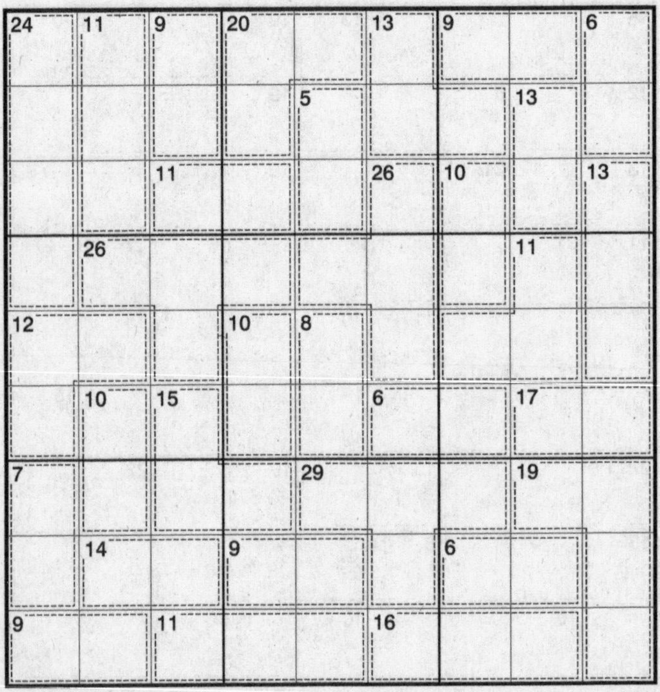

1 HOUR 20 MINUTES

TIME TAKEN..........................

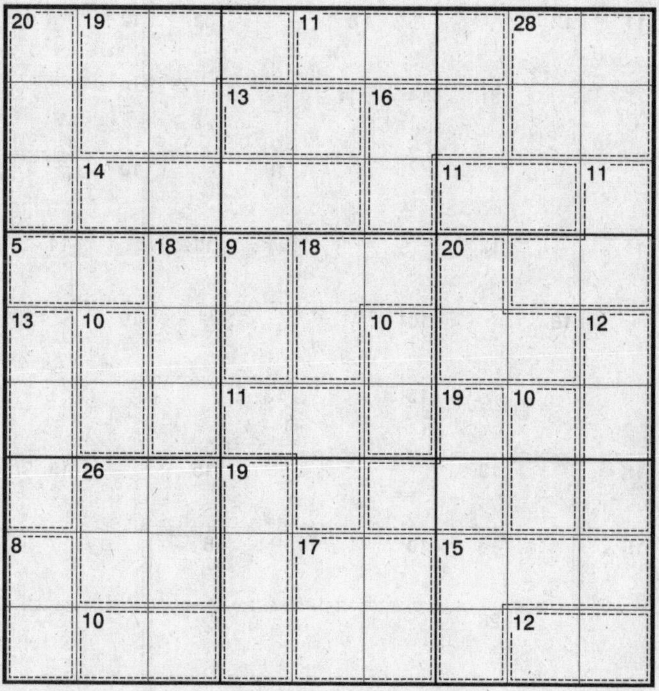

1 HOUR 20 MINUTES

TIME TAKEN............................

Deadly

92

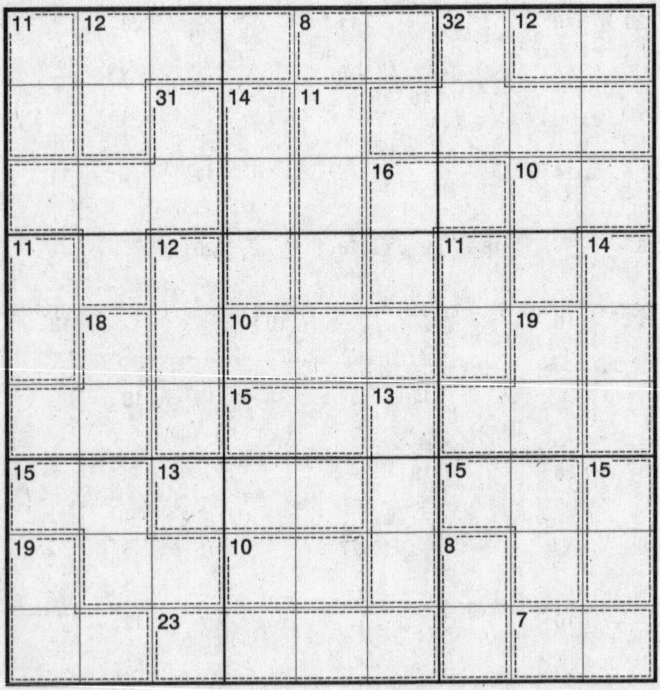

1 HOUR 20 MINUTES

TIME TAKEN..........................

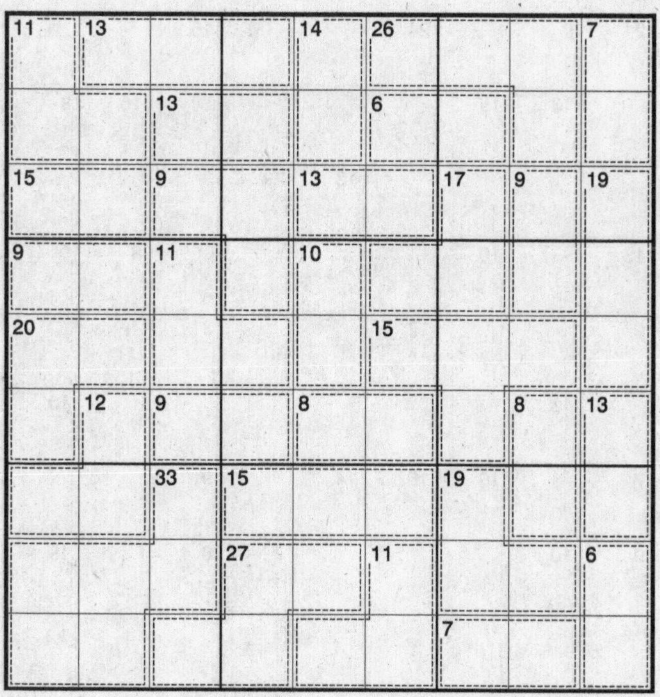

1 HOUR 20 MINUTES

TIME TAKEN............................

Deadly

94

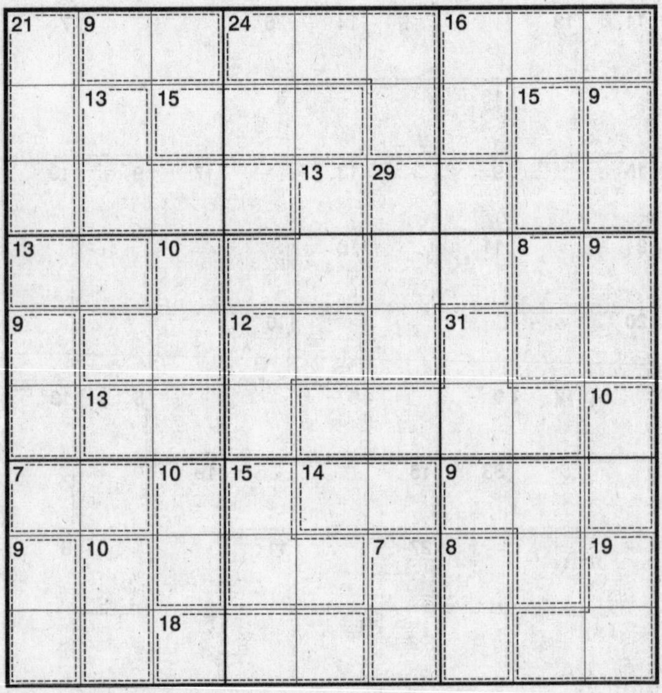

1 HOUR 20 MINUTES

TIME TAKEN..........................

95

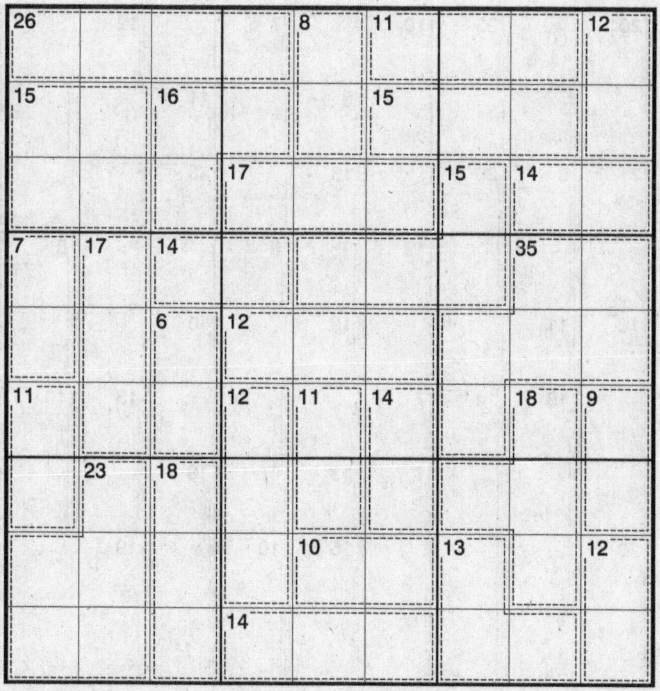

1 HOUR 25 MINUTES

TIME TAKEN..........................

Deadly

96

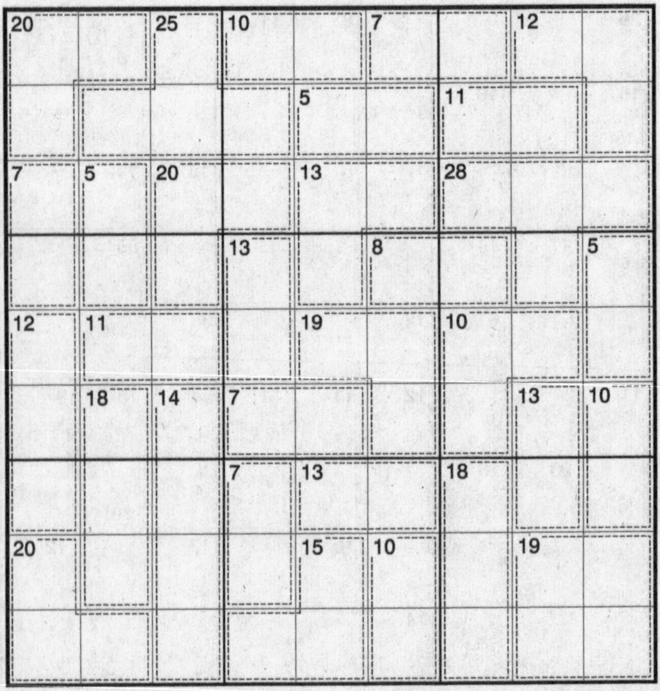

1 HOUR 25 MINUTES

TIME TAKEN..........................

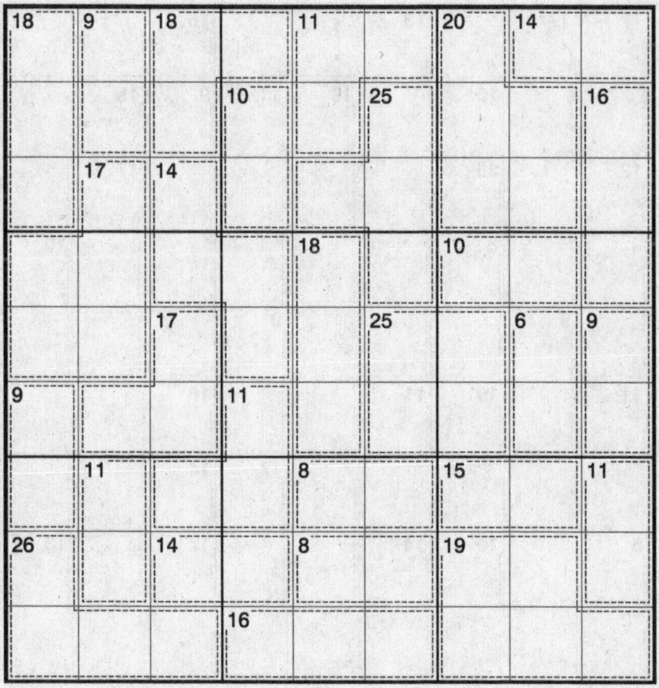

1 HOUR 25 MINUTES

TIME TAKEN.............................

Deadly

98

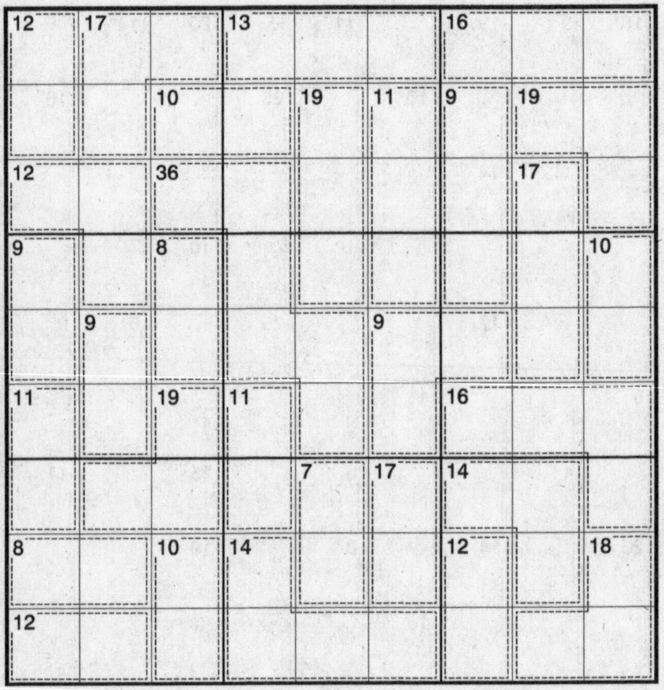

1 HOUR 25 MINUTES

TIME TAKEN.............................

1 HOUR 25 MINUTES

TIME TAKEN..........................

Deadly

100

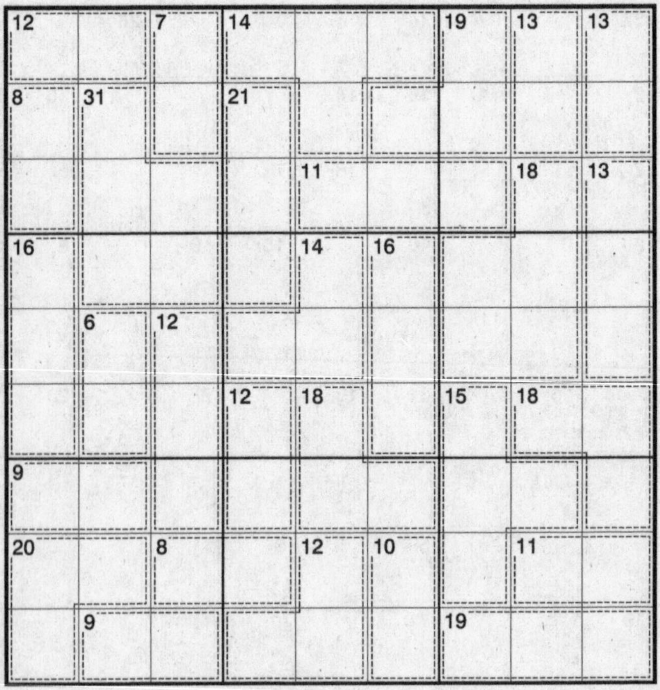

1 HOUR 25 MINUTES

TIME TAKEN............................

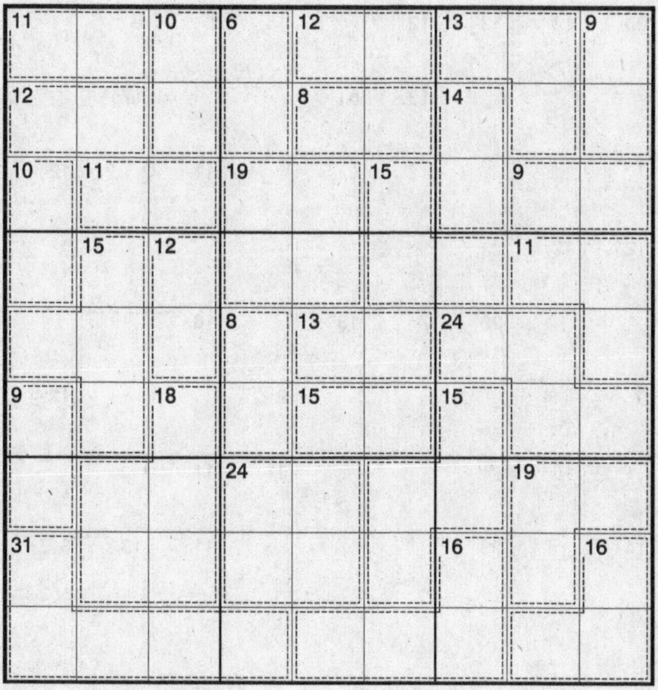

1 HOUR 30 MINUTES

TIME TAKEN..........................

102

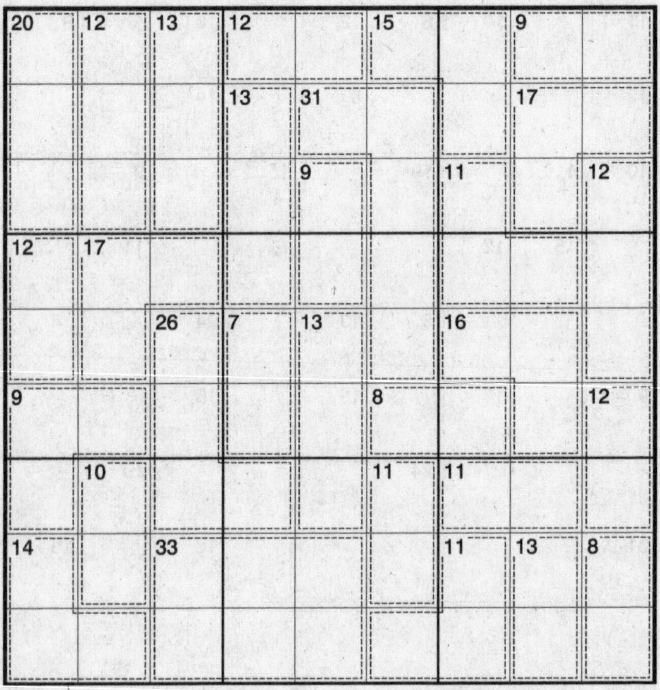

1 HOUR 30 MINUTES

TIME TAKEN..........................

103

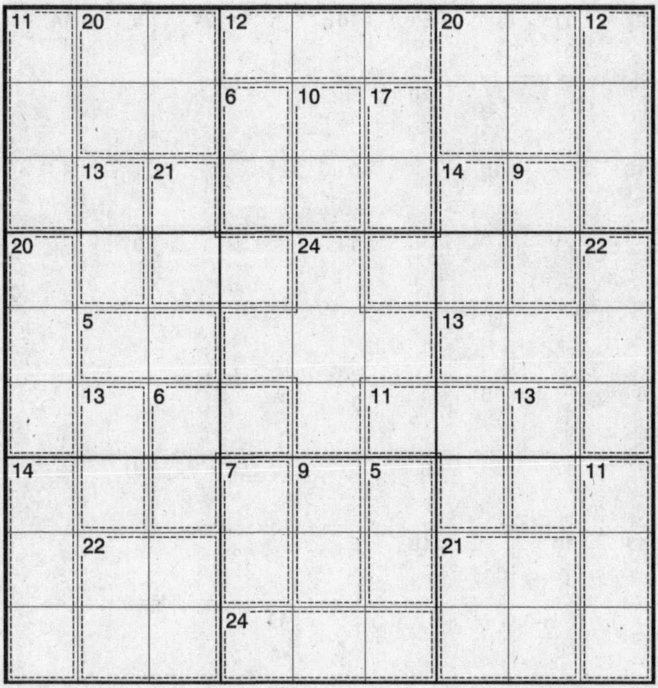

1 HOUR 40 MINUTES

TIME TAKEN..........................

Extra Deadly

104

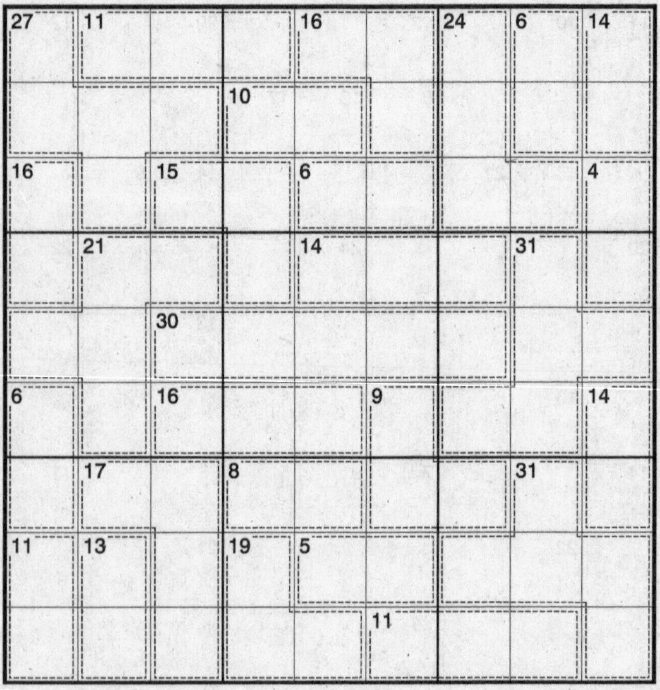

1 HOUR 40 MINUTES

TIME TAKEN.........................

105

1 HOUR 40 MINUTES

TIME TAKEN..........................

Extra Deadly

106

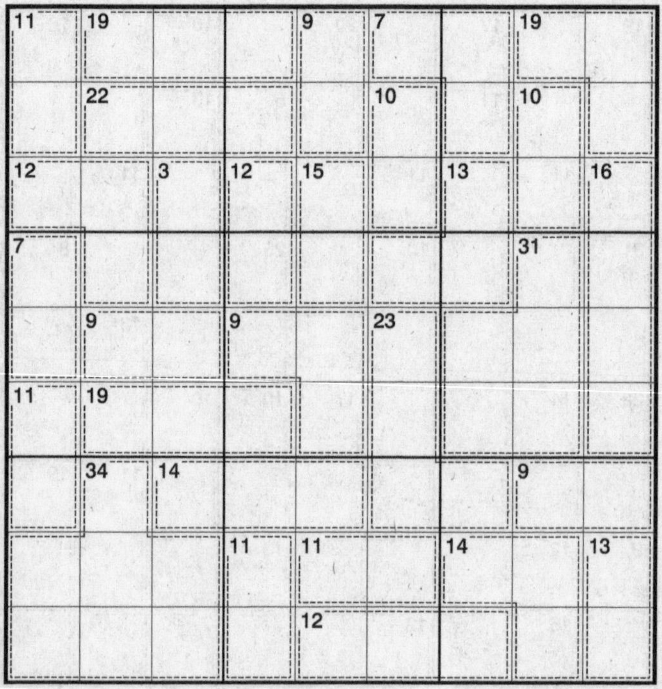

1 HOUR 40 MINUTES

TIME TAKEN..........................

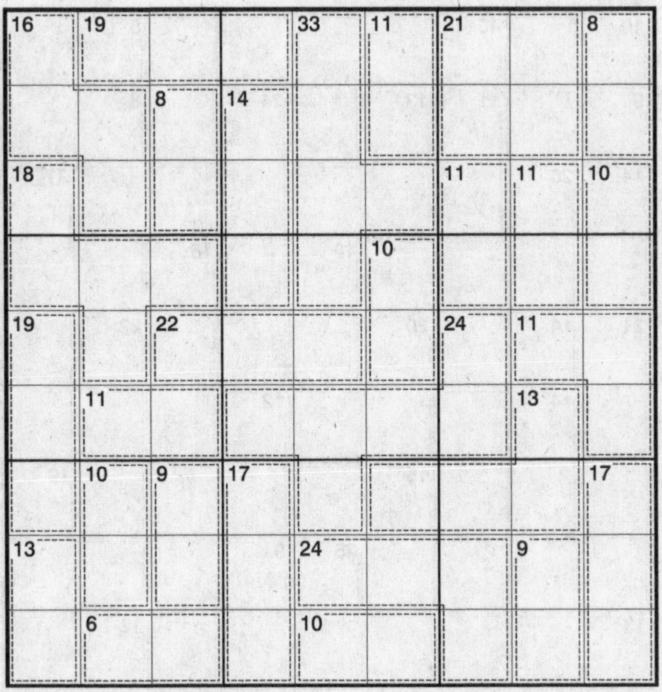

1 HOUR 50 MINUTES

TIME TAKEN..........................

108

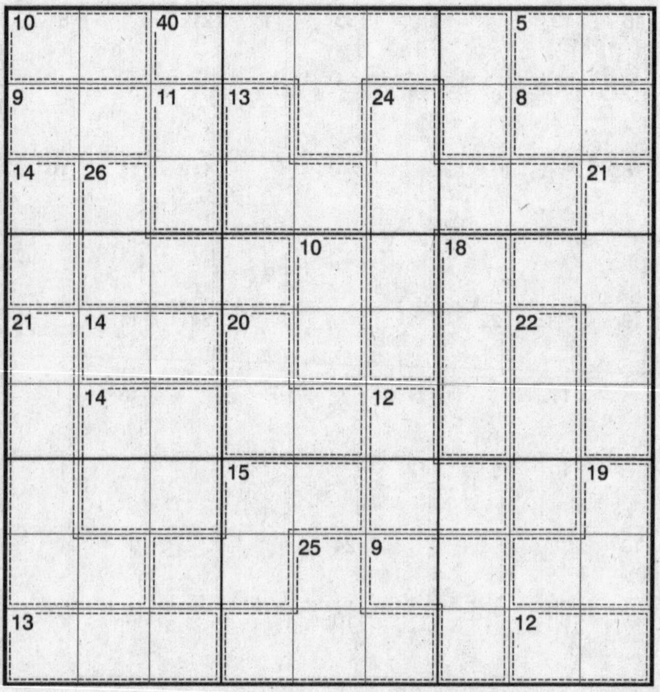

1 HOUR 50 MINUTES

TIME TAKEN..........................

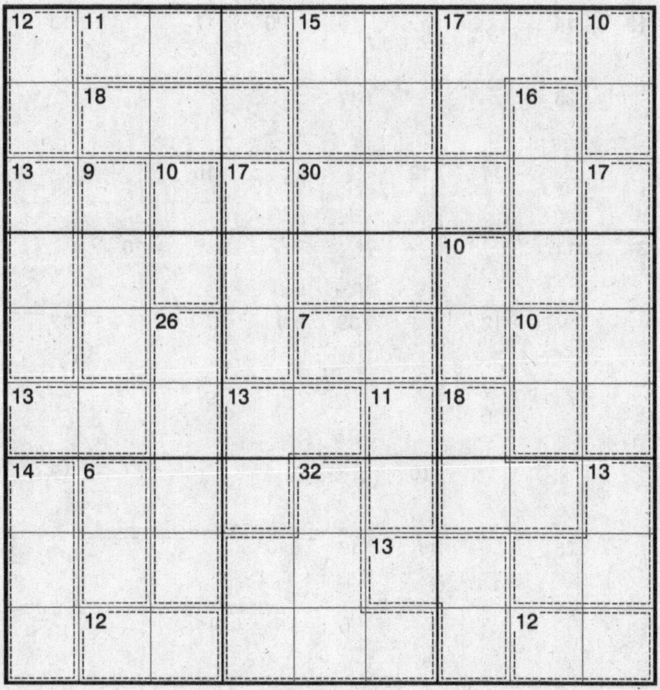

1 HOUR 55 MINUTES

TIME TAKEN..........................

110

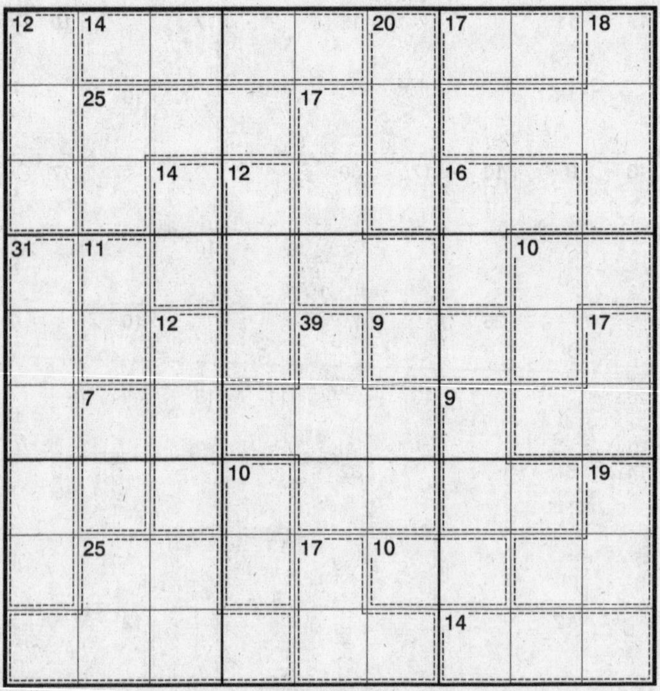

2 HOURS

TIME TAKEN.........................

111

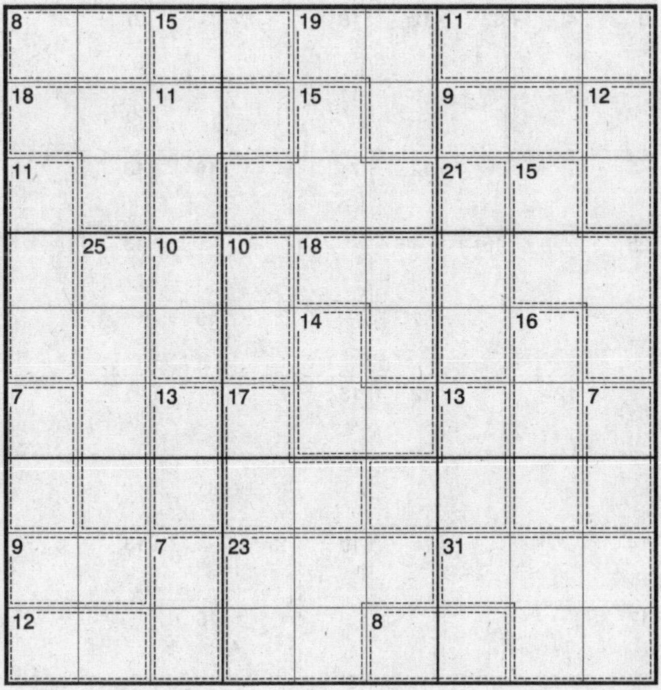

2 HOURS 5 MINUTES

TIME TAKEN..........................

Extra Deadly

112

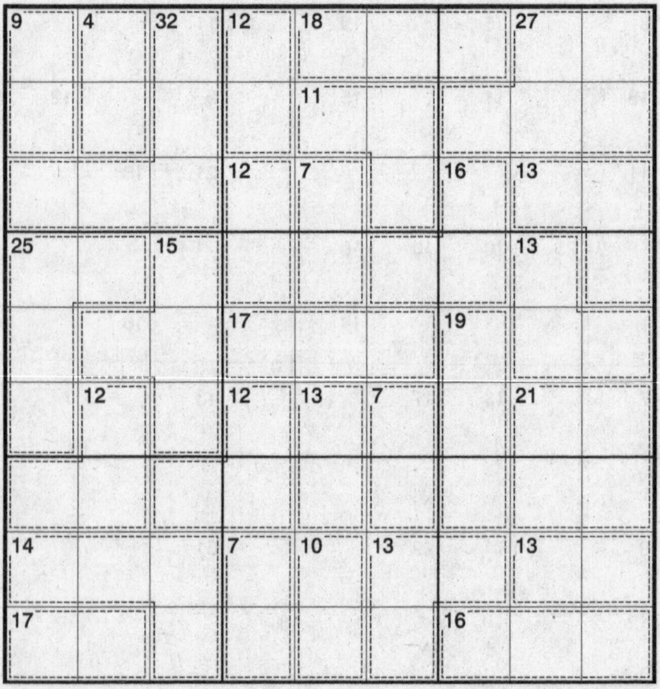

2 HOURS 10 MINUTES

TIME TAKEN..........................

113

2 HOURS 10 MINUTES

TIME TAKEN..........................

Extra Deadly

114

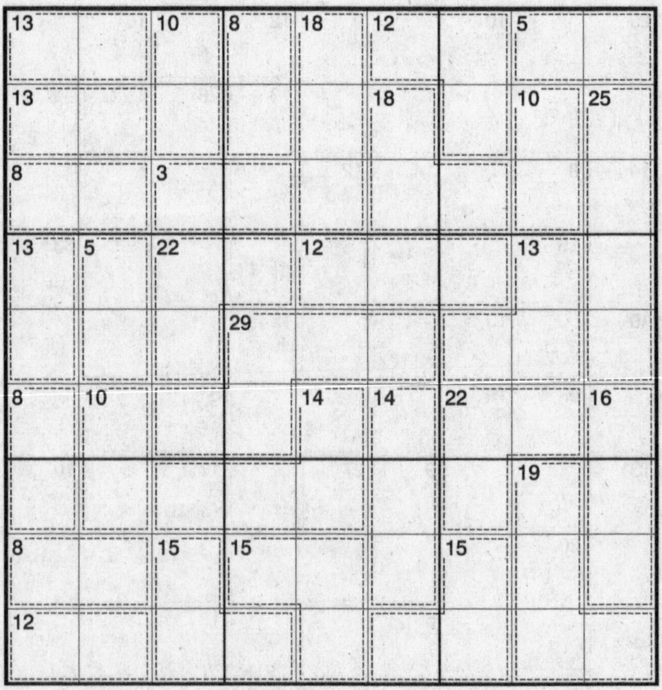

2 HOURS 15 MINUTES

TIME TAKEN.........................

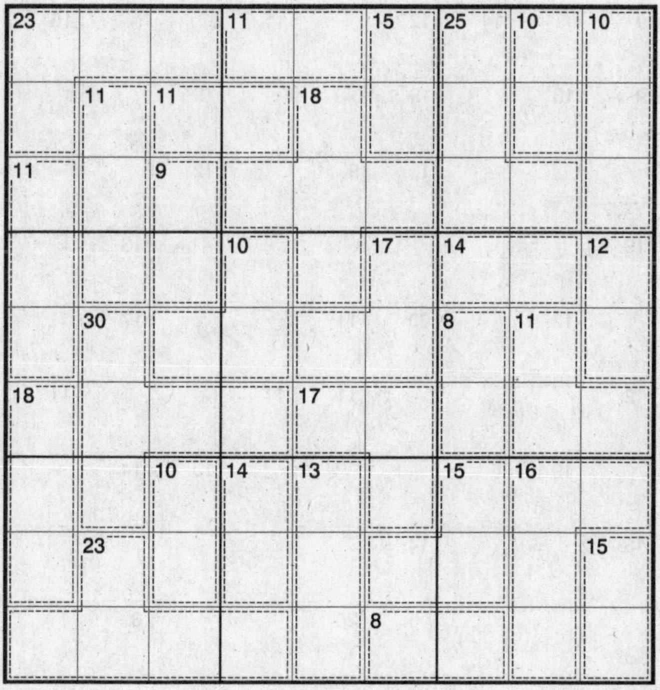

2 HOURS 15 MINUTES

TIME TAKEN.............................

116

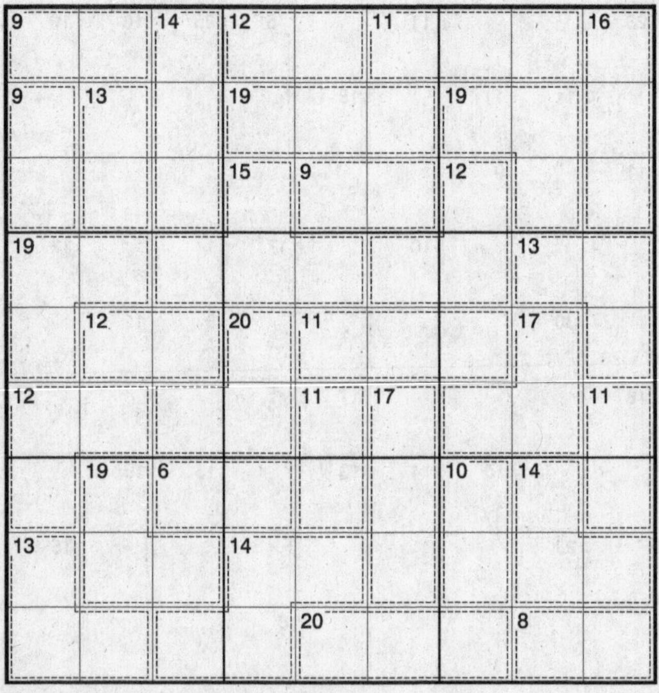

2 HOURS 20 MINUTES

TIME TAKEN.........................

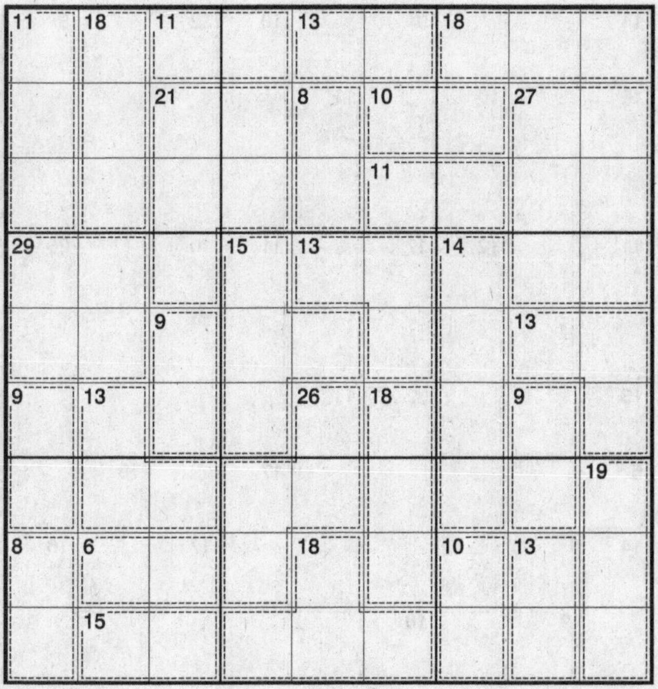

2 HOURS 20 MINUTES

TIME TAKEN..............................

Extra Deadly

118

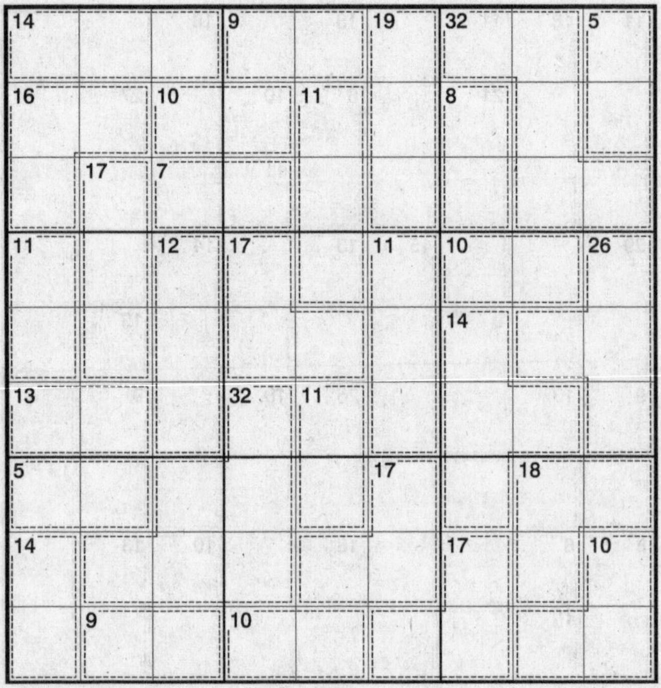

2 HOURS 25 MINUTES

TIME TAKEN.........................

2 HOURS 35 MINUTES

TIME TAKEN..........................

Extra Deadly

120

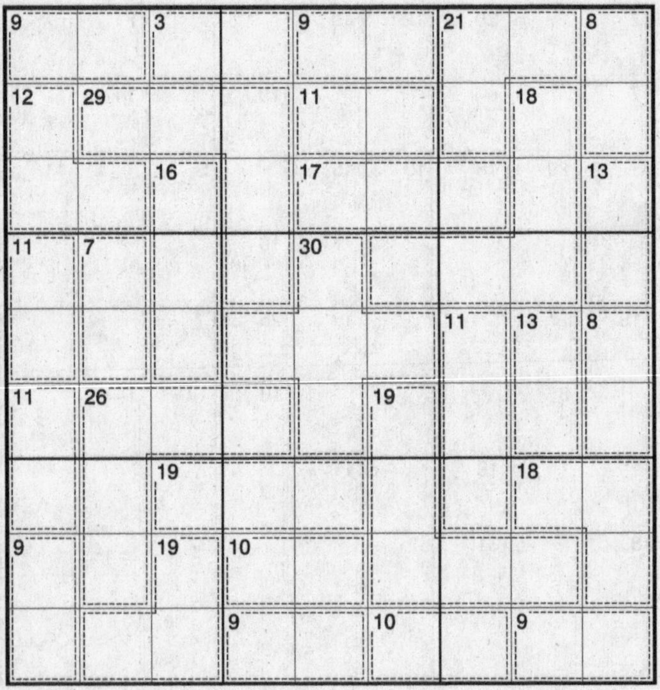

2 HOURS 35 MINUTES

TIME TAKEN..........................

PUZZLES
Book Two

1

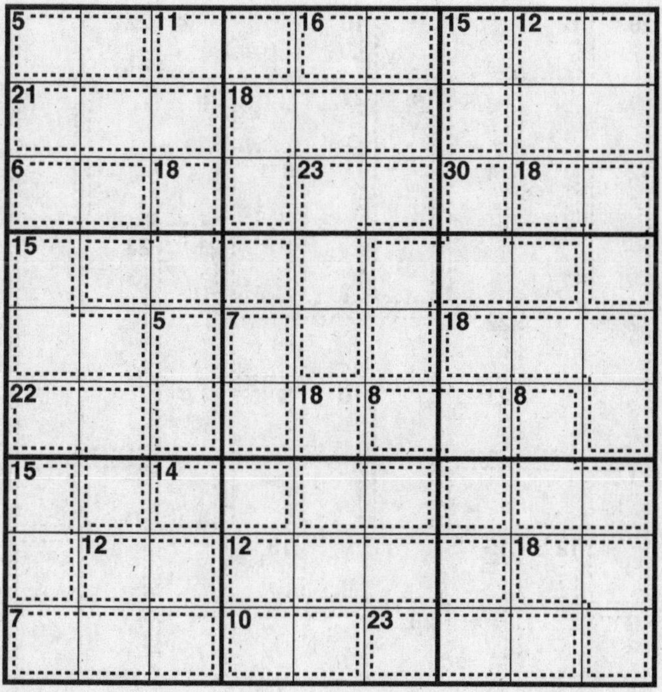

1 HOUR

TIME TAKEN.........................

Deadly

2

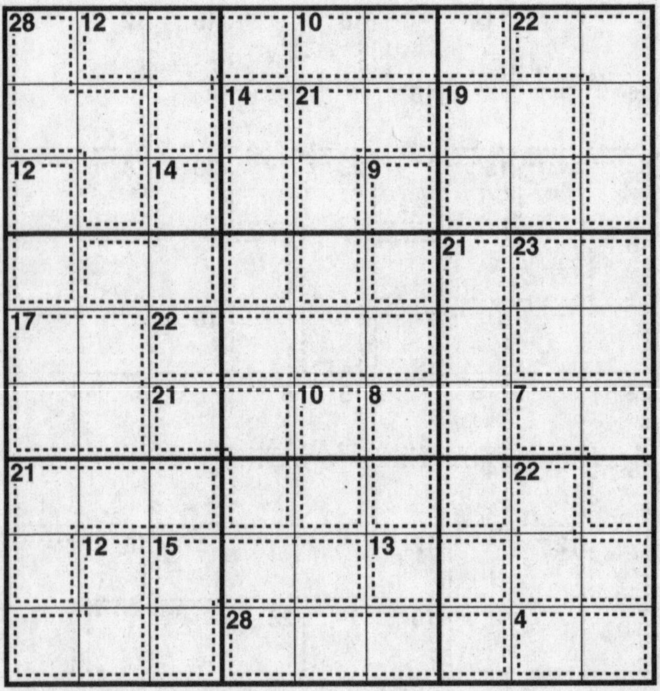

1 HOUR

TIME TAKEN..........................

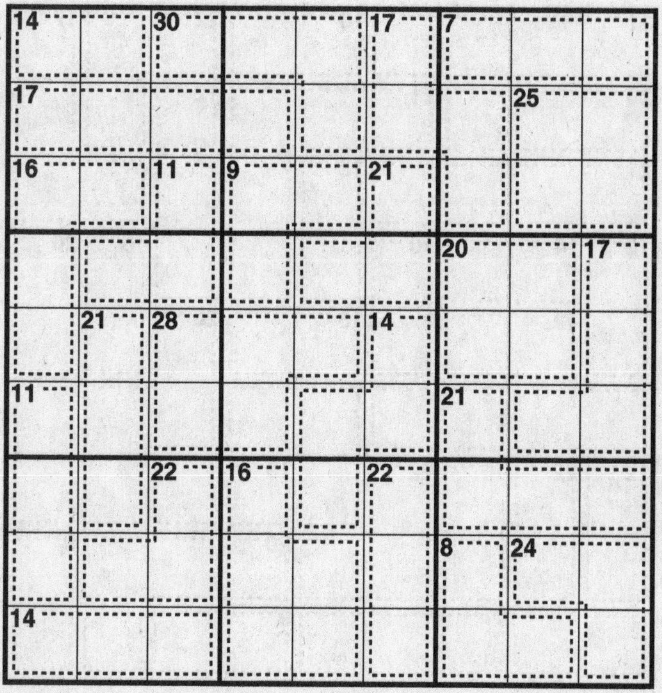

1 HOUR

TIME TAKEN.............................

4

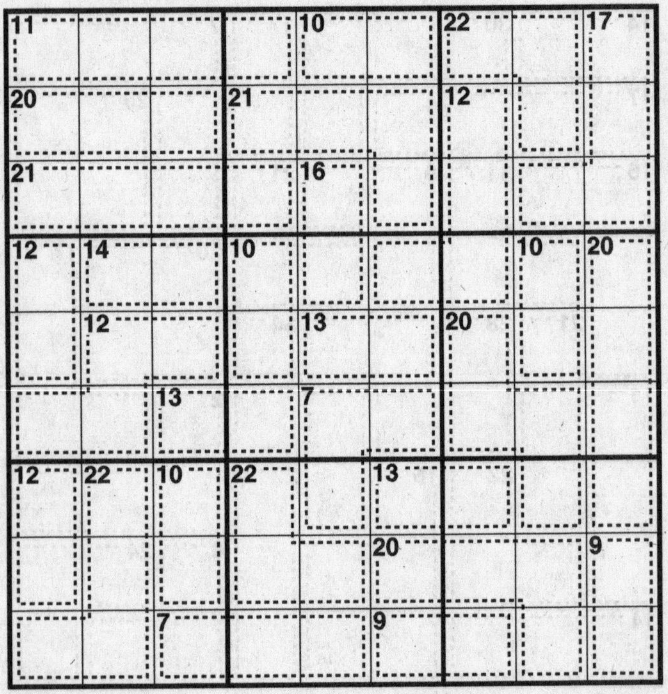

1 HOUR

TIME TAKEN..........................

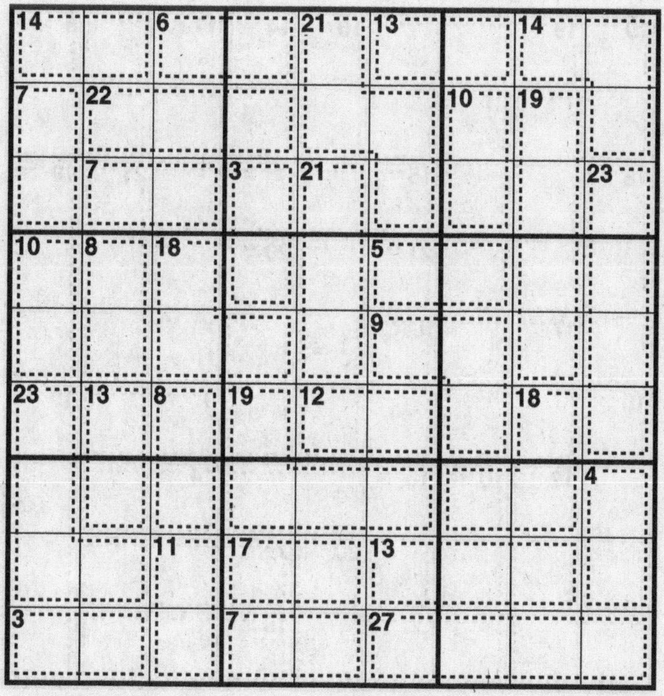

1 HOUR

TIME TAKEN..........................

Deadly

6

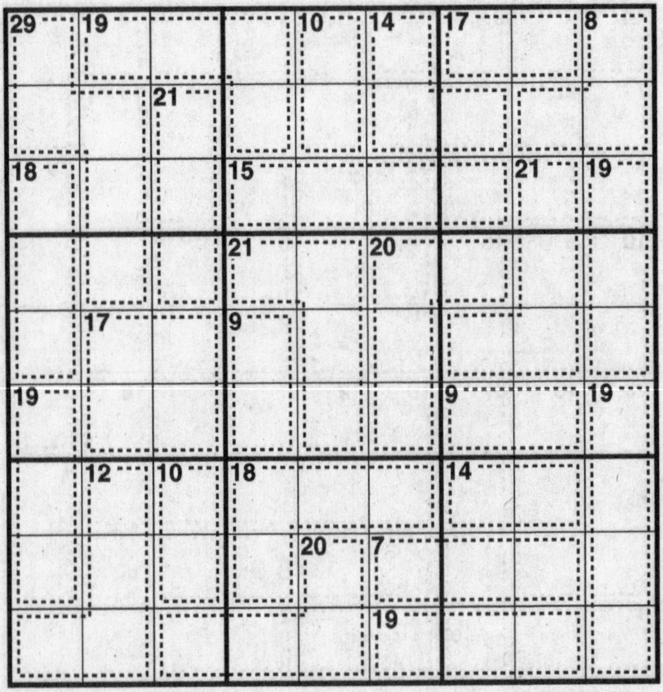

1 HOUR 5 MINUTES

TIME TAKEN..........................

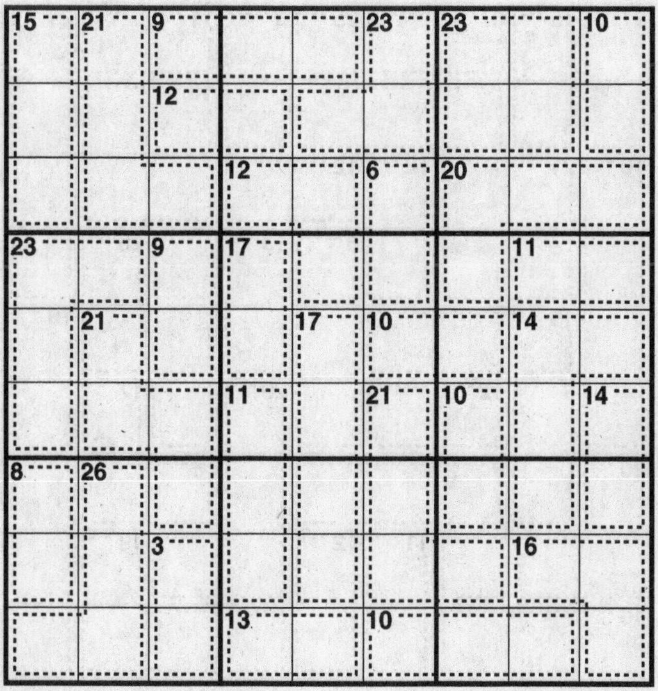

1 HOUR 5 MINUTES

TIME TAKEN...........................

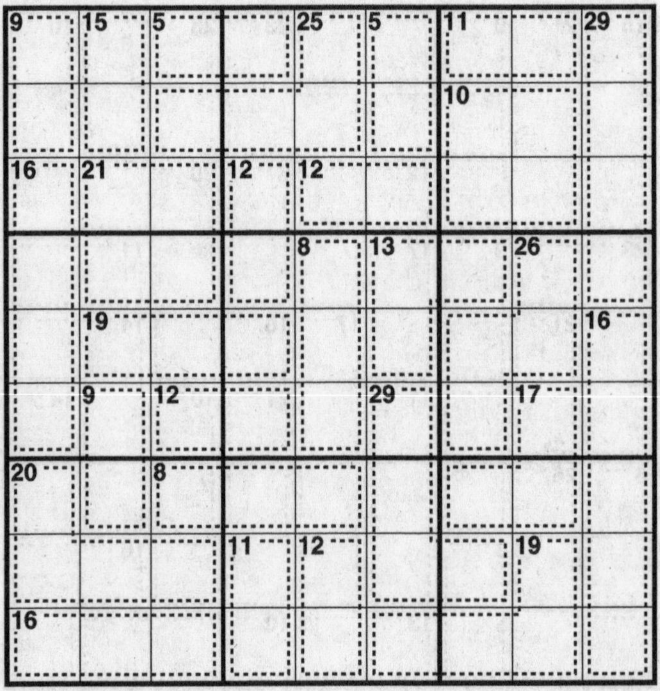

1 HOUR 5 MINUTES

TIME TAKEN............................

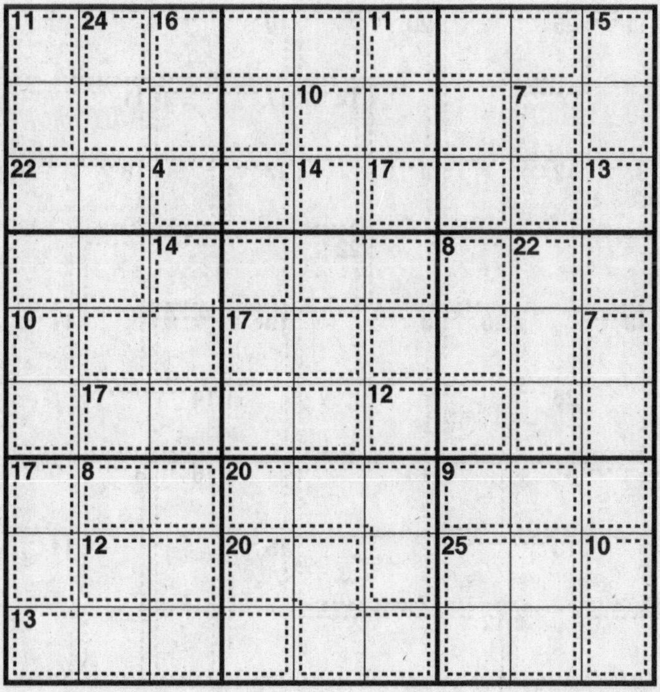

1 HOUR 5 MINUTES

TIME TAKEN.........................

10

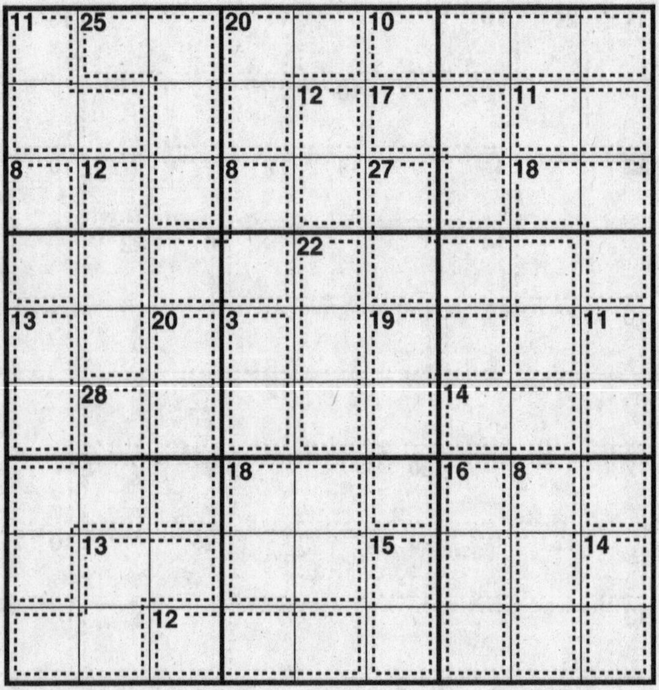

1 HOUR 5 MINUTES

TIME TAKEN..........................

11

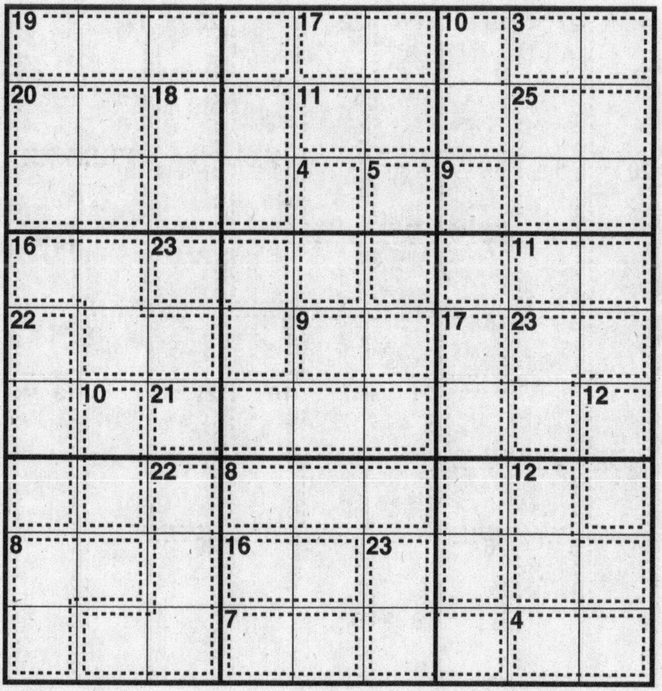

1 HOUR 5 MINUTES

TIME TAKEN..........................

Deadly

12

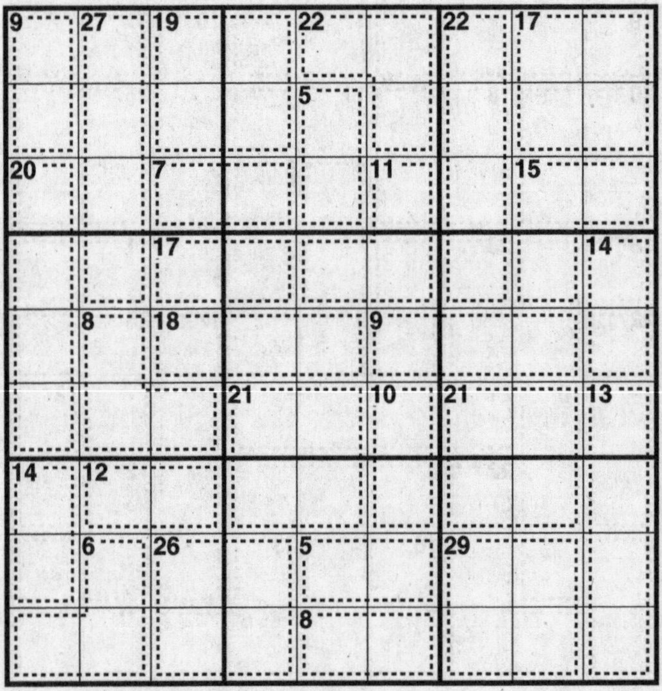

1 HOUR 5 MINUTES

TIME TAKEN.........................

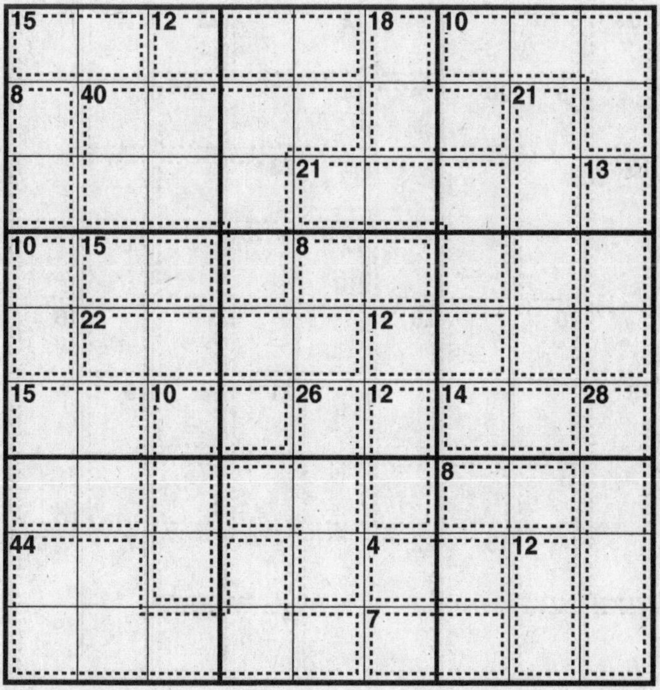

1 HOUR 5 MINUTES

TIME TAKEN..........................

14

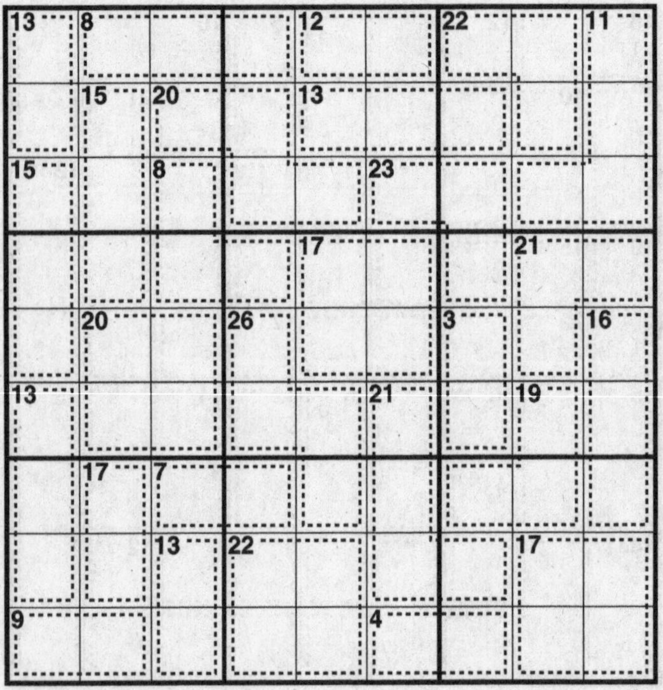

1 HOUR 5 MINUTES

TIME TAKEN.........................

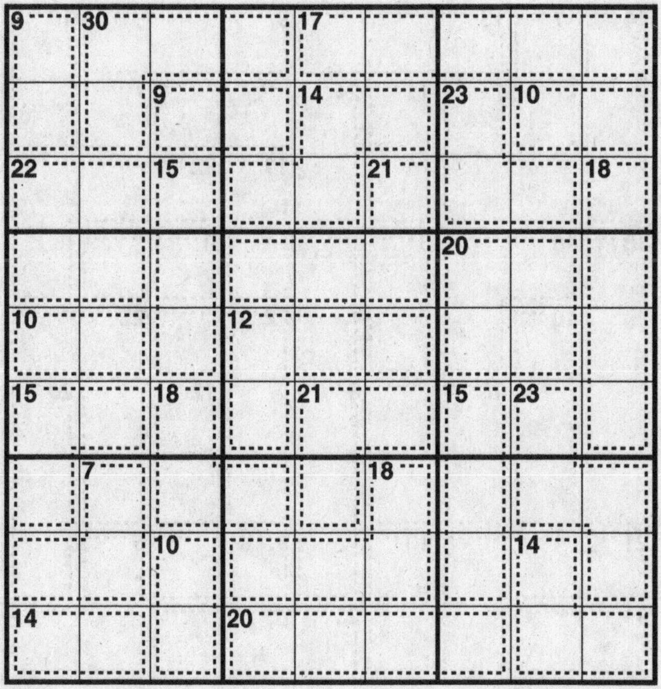

1 HOUR 5 MINUTES

TIME TAKEN..........................

16

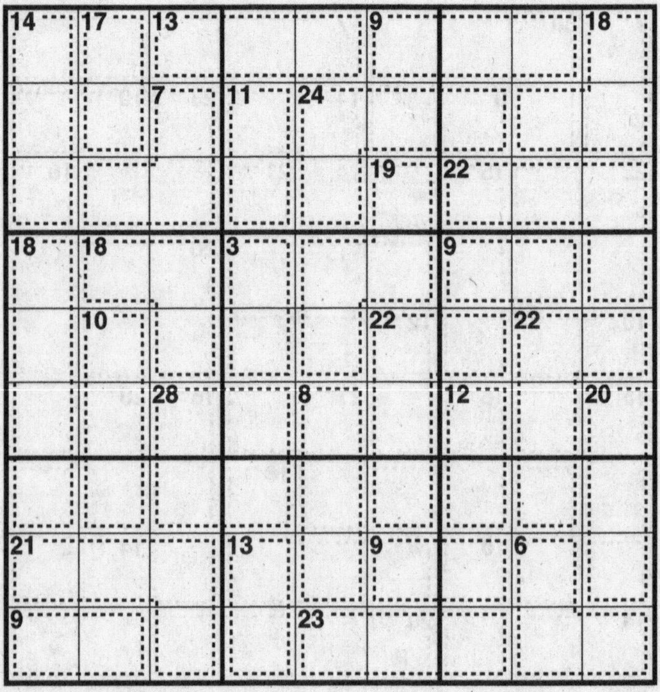

1 HOUR 10 MINUTES

TIME TAKEN.........................

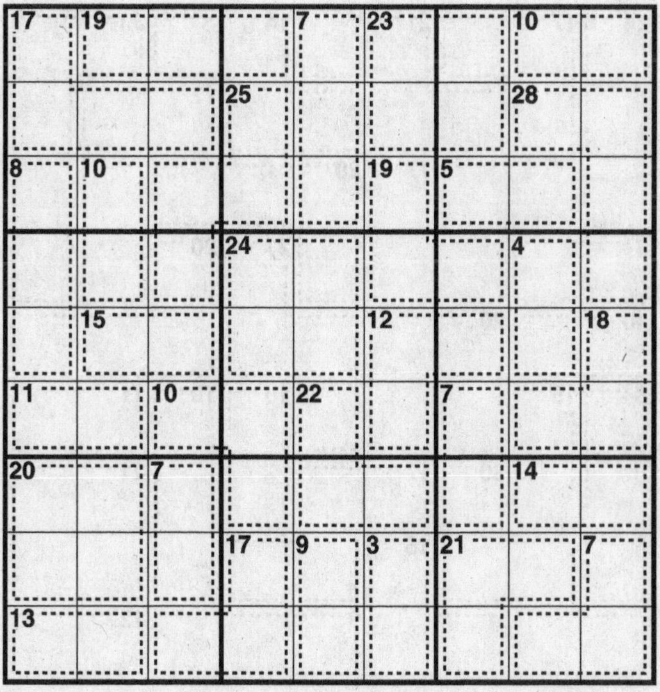

1 HOUR 10 MINUTES

TIME TAKEN..........................

Deadly

18

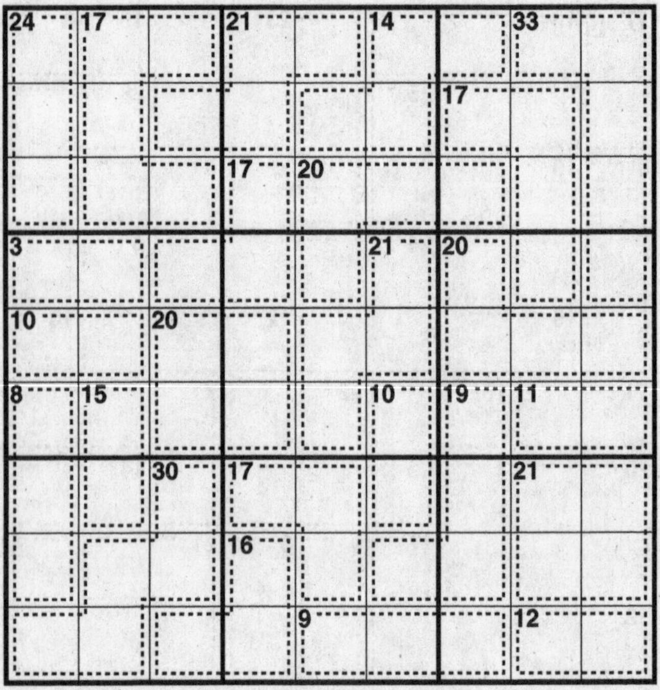

1 HOUR 10 MINUTES

TIME TAKEN..........................

19

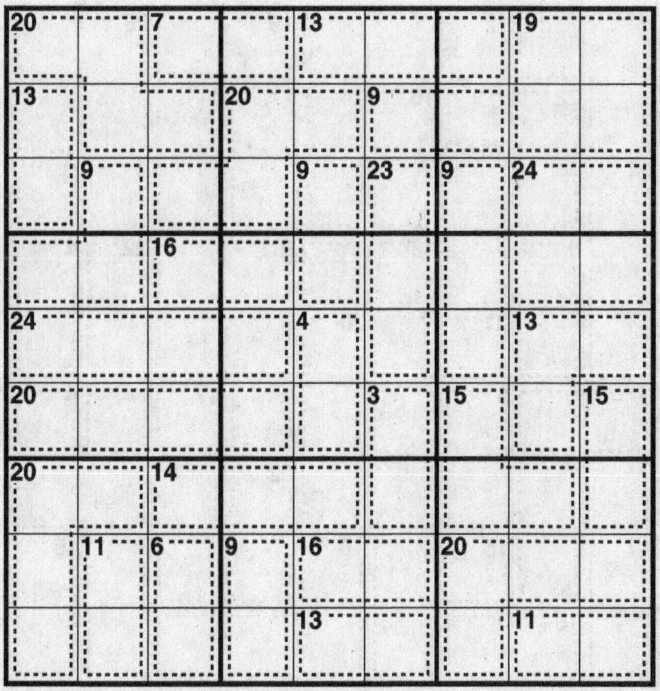

1 HOUR 10 MINUTES

TIME TAKEN.........................

Deadly

20

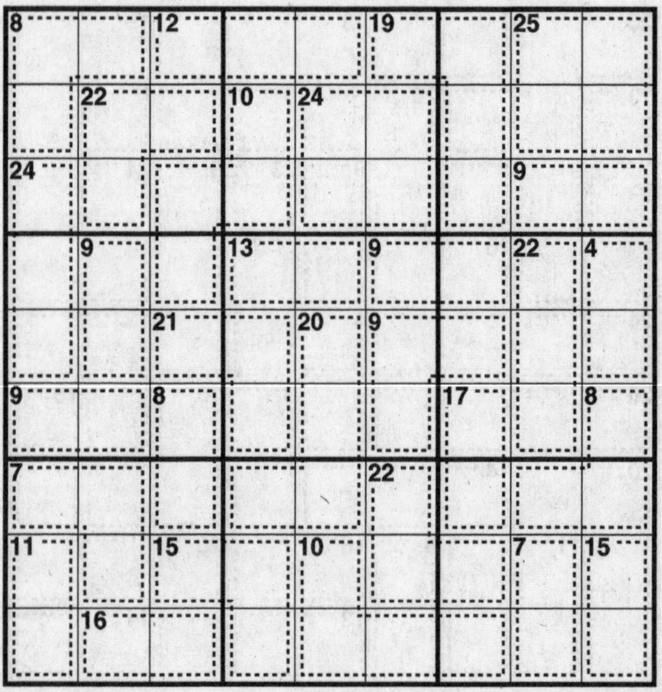

1 HOUR 10 MINUTES

TIME TAKEN..........................

21

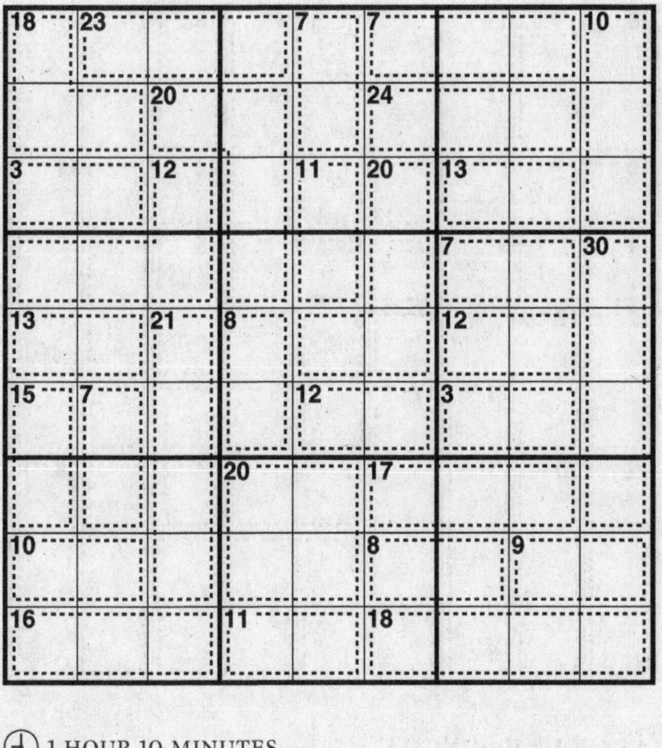

1 HOUR 10 MINUTES

TIME TAKEN..........................

Deadly

22

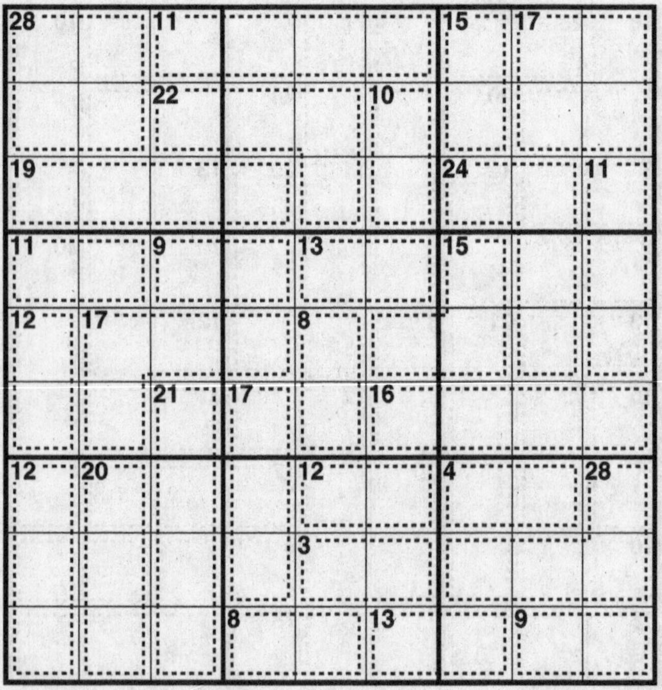

1 HOUR 10 MINUTES

TIME TAKEN..........................

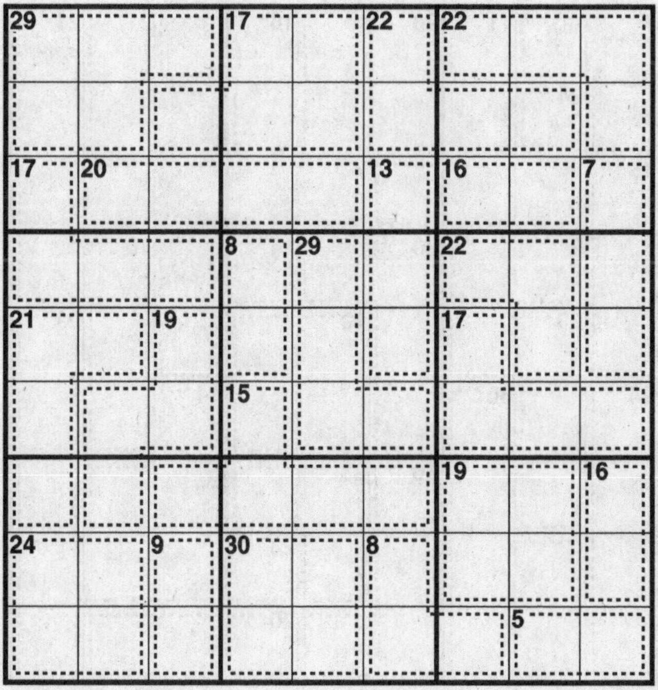

1 HOUR 10 MINUTES

TIME TAKEN..............................

Deadly

24

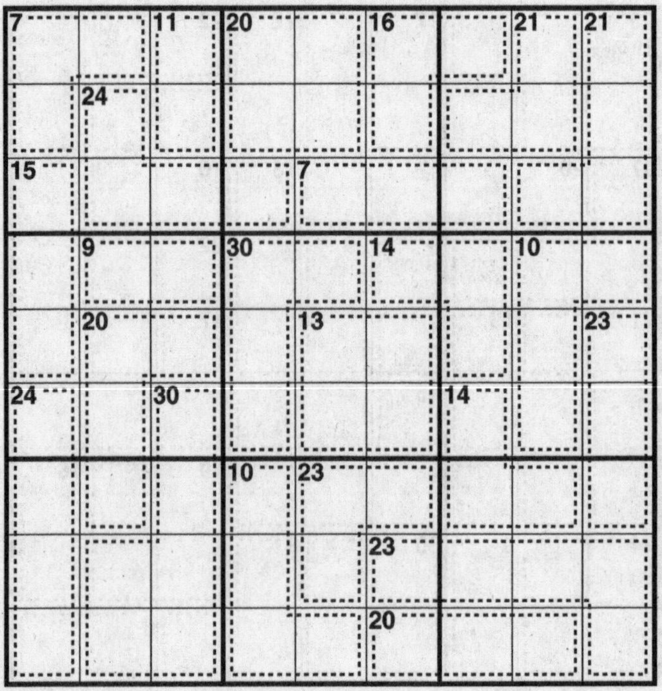

1 HOUR 10 MINUTES

TIME TAKEN..............................

25

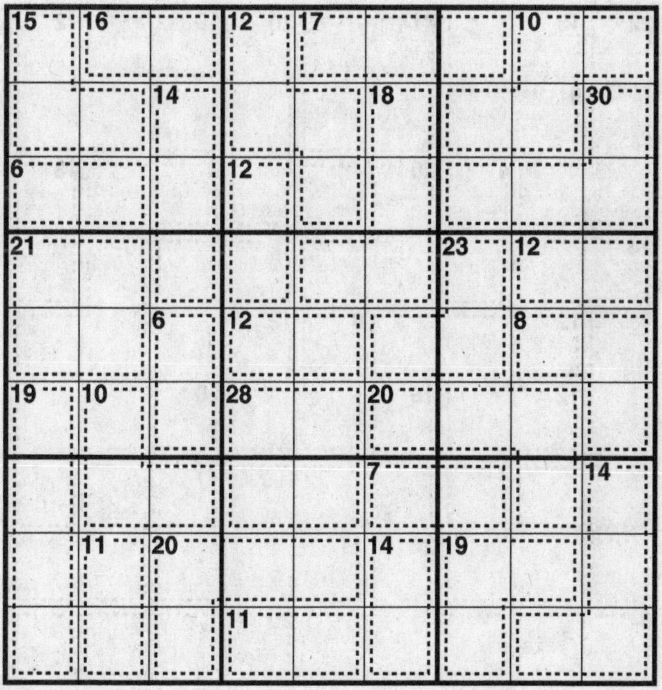

1 HOUR 10 MINUTES

TIME TAKEN.........................

Deadly

26

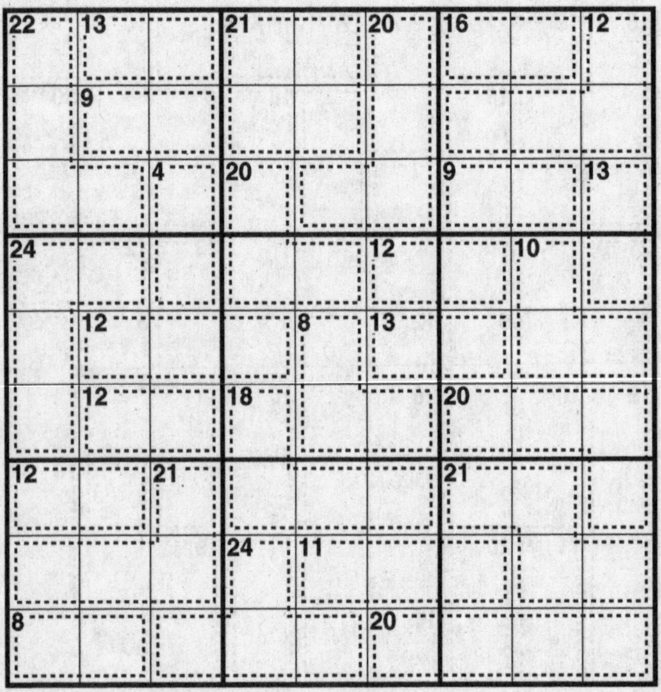

1 HOUR 10 MINUTES

TIME TAKEN............................

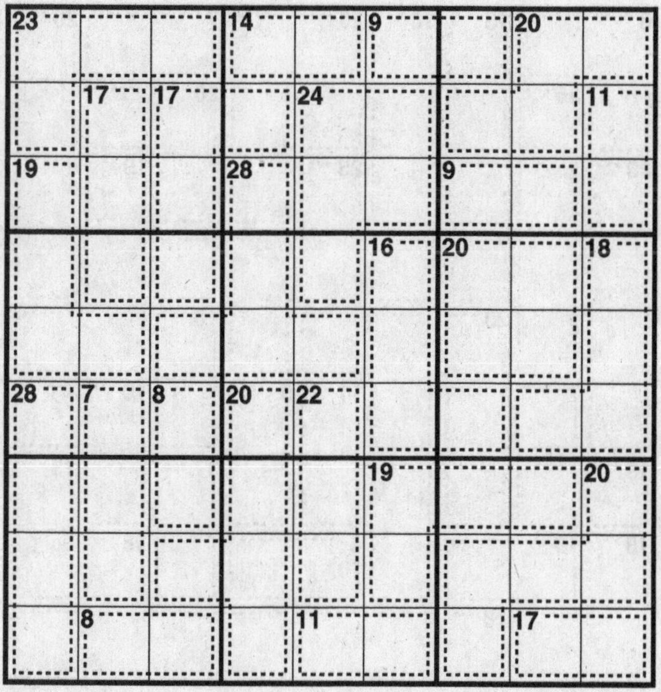

1 HOUR 10 MINUTES

TIME TAKEN..........................

Deadly

28

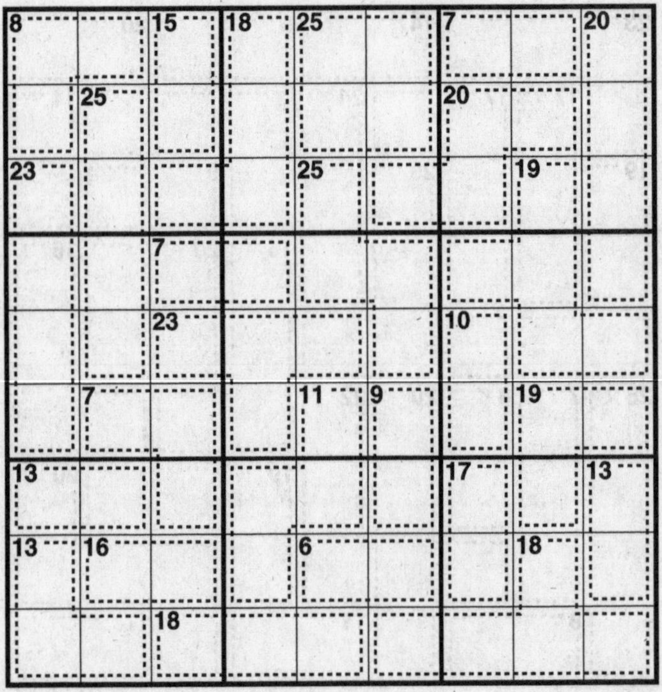

1 HOUR 10 MINUTES

TIME TAKEN........................

29

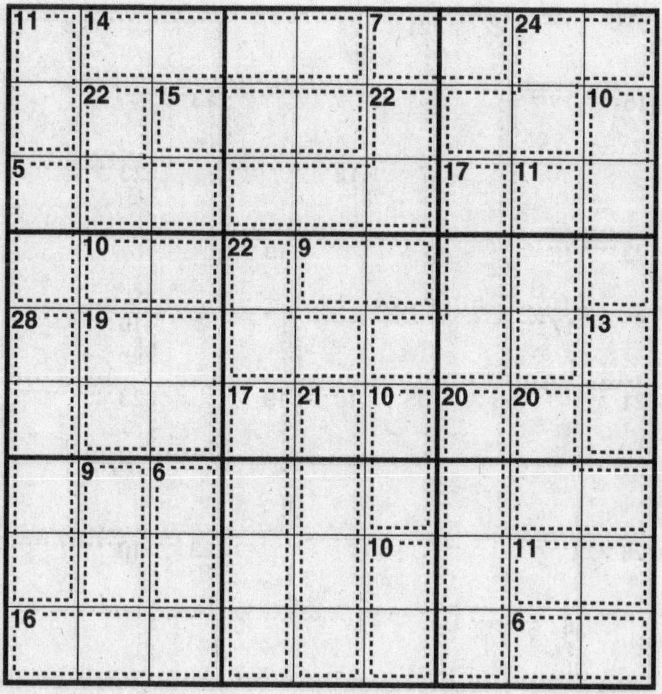

1 HOUR 10 MINUTES

TIME TAKEN..........................

Deadly

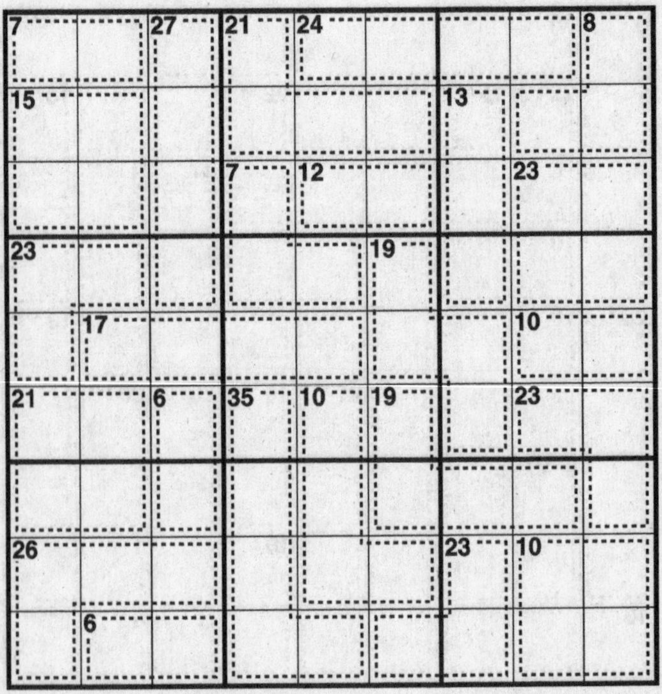

1 HOUR 15 MINUTES

TIME TAKEN............................

31

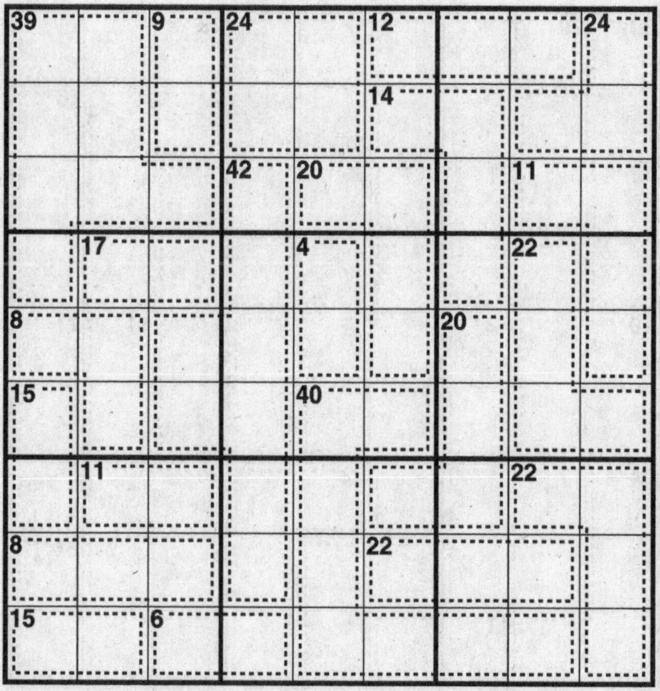

1 HOUR 15 MINUTES

TIME TAKEN..........................

Deadly

32

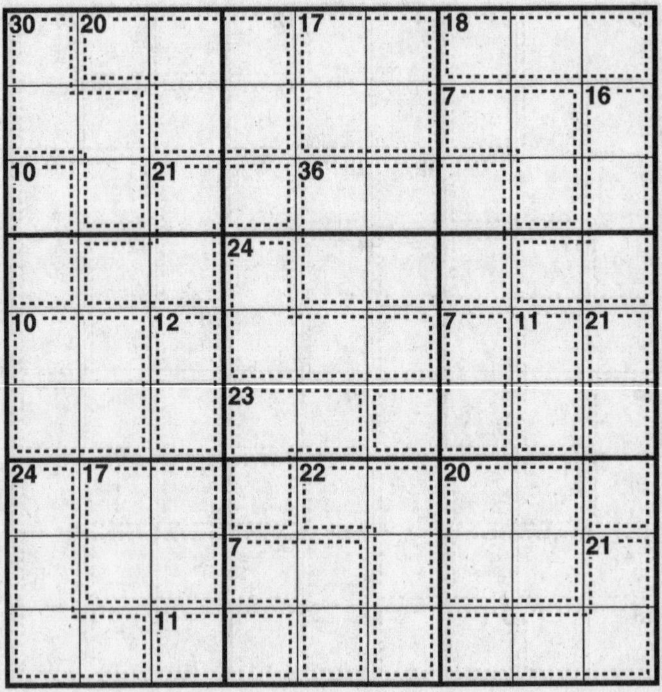

1 HOUR 15 MINUTES

TIME TAKEN..........................

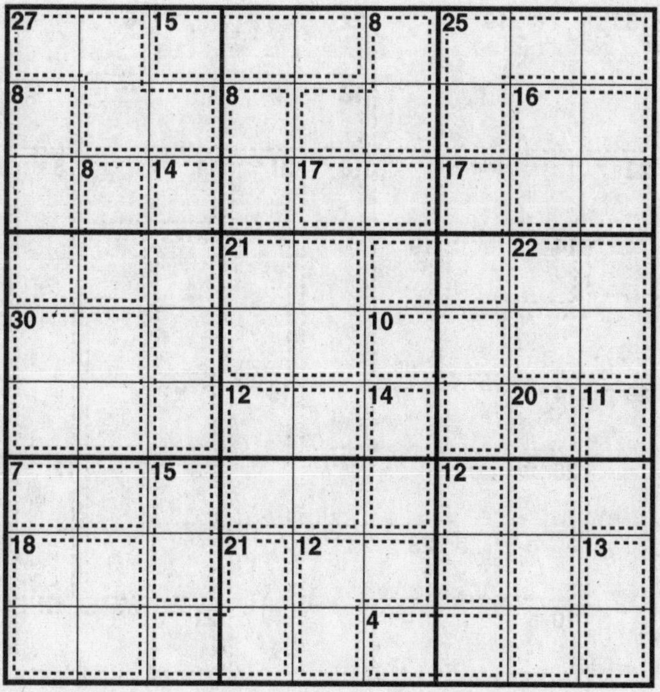

1 HOUR 15 MINUTES

TIME TAKEN..........................

34

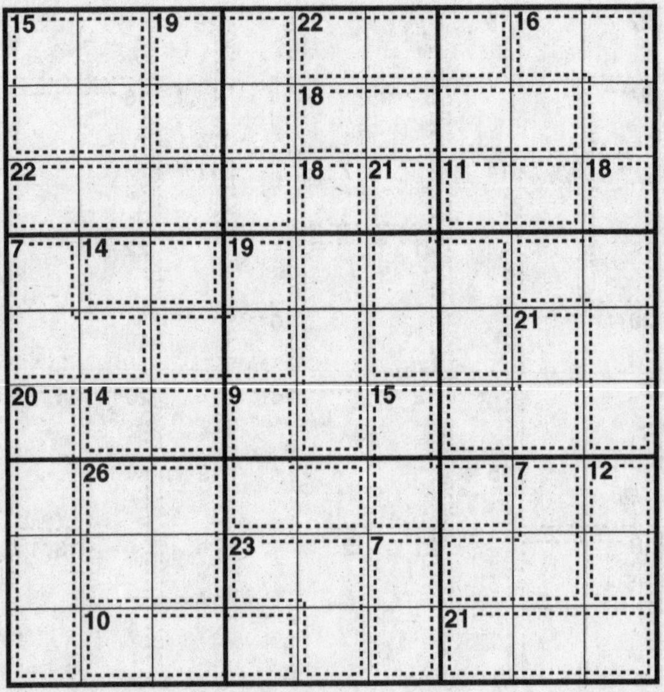

1 HOUR 15 MINUTES

TIME TAKEN..........................

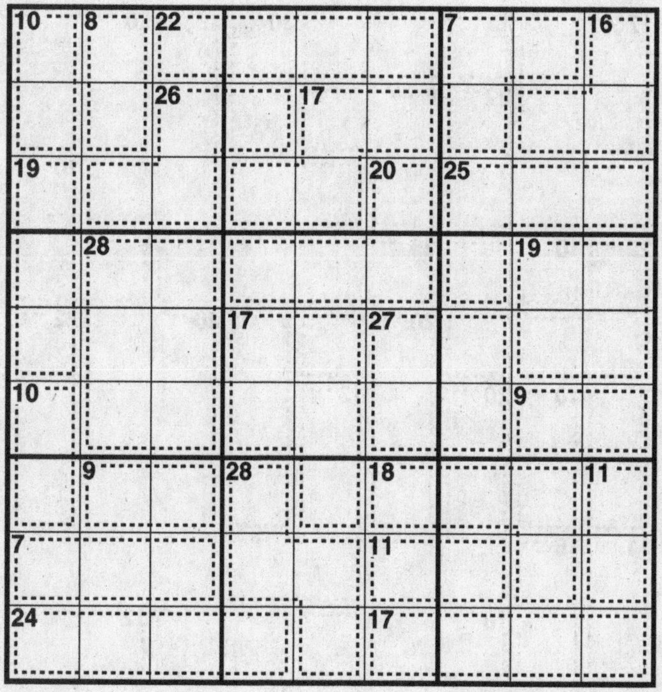

1 HOUR 15 MINUTES

TIME TAKEN............................

36

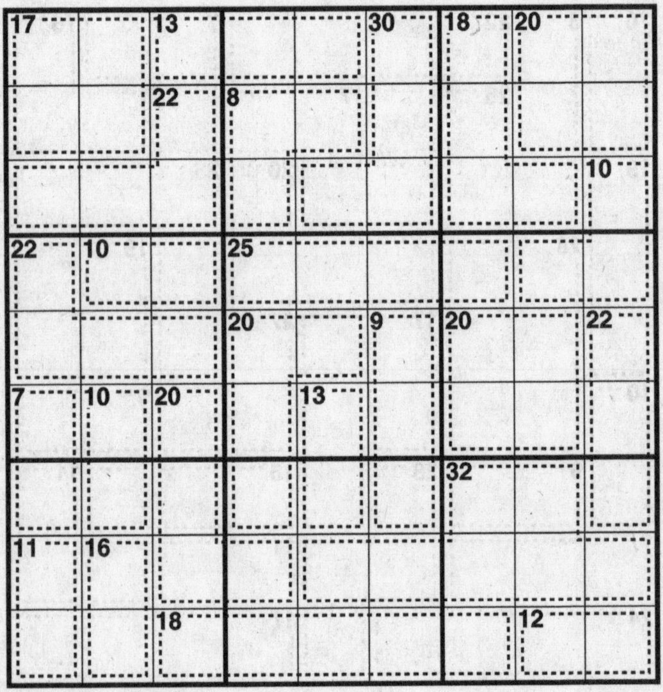

1 HOUR 20 MINUTES

TIME TAKEN.........................

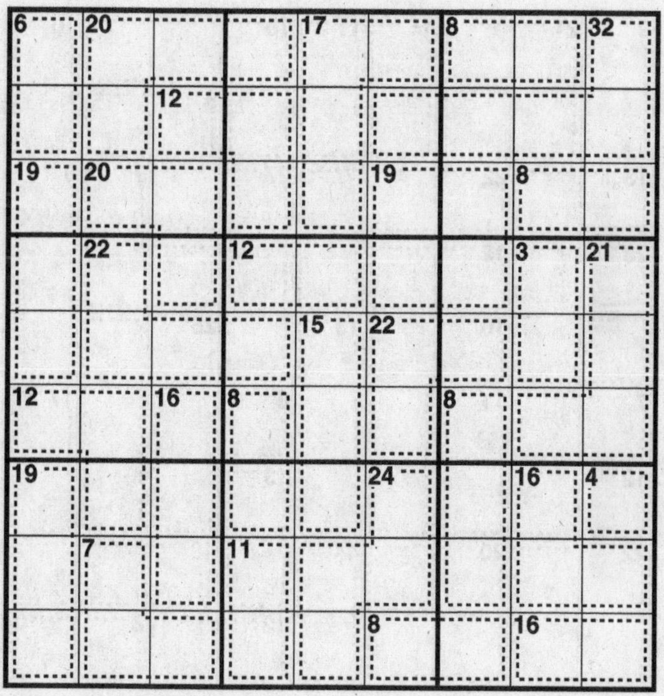

1 HOUR 20 MINUTES

TIME TAKEN............................

38

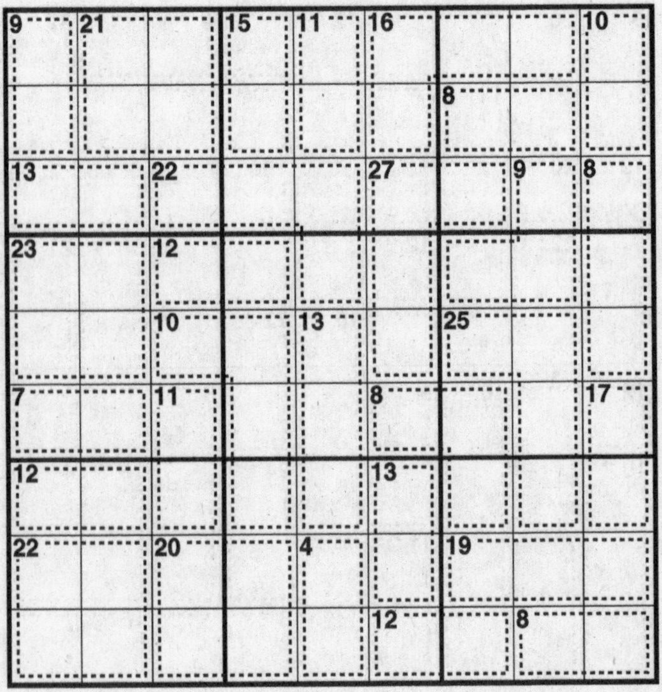

1 HOUR 20 MINUTES

TIME TAKEN............................

39

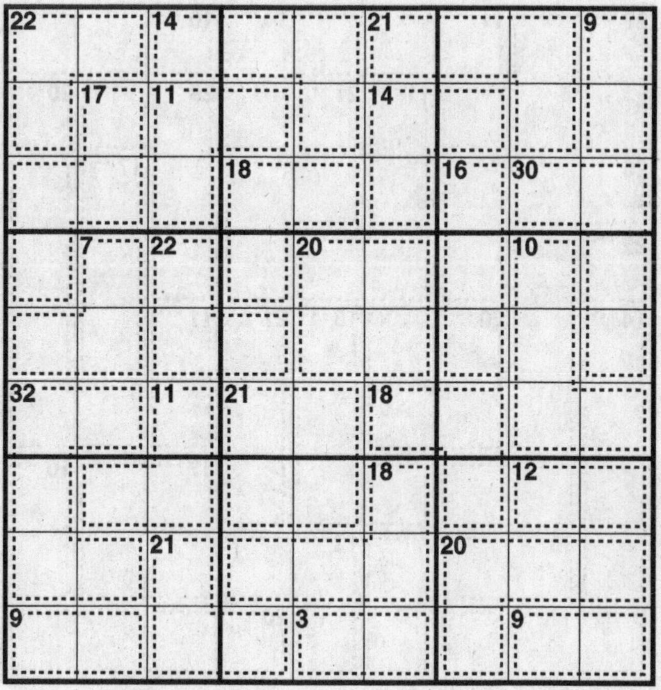

1 HOUR 20 MINUTES

TIME TAKEN.........................

Deadly

40

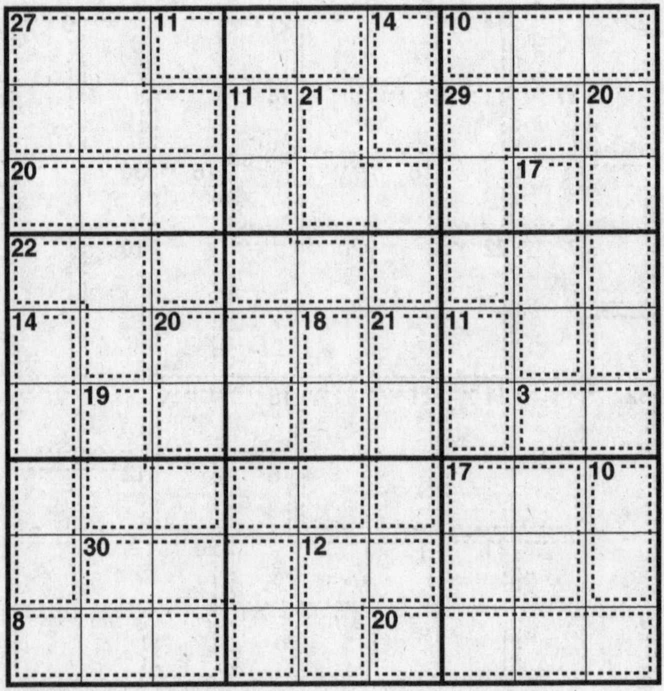

1 HOUR 20 MINUTES

TIME TAKEN..........................

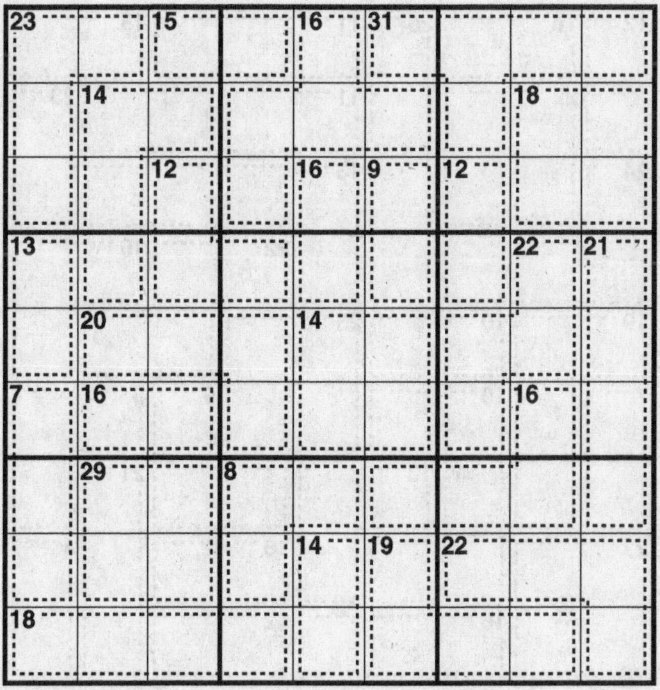

1 HOUR 20 MINUTES

TIME TAKEN.........................

Deadly

42

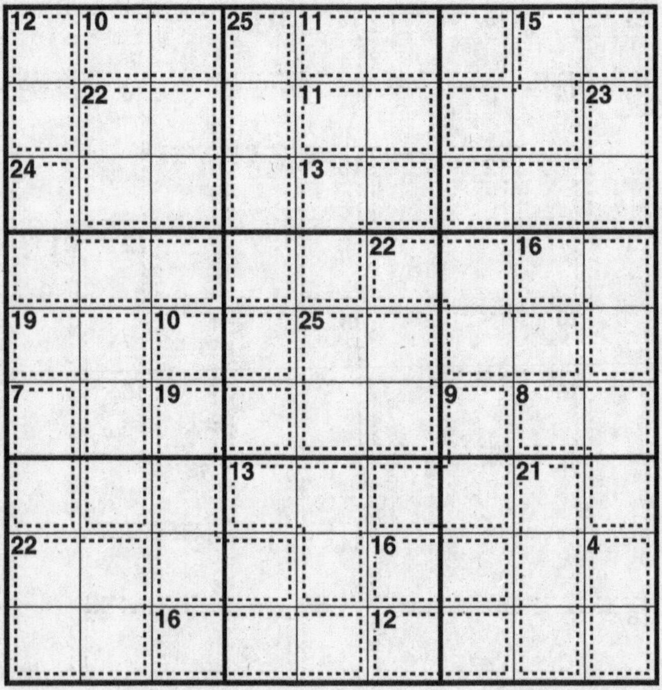

1 HOUR 20 MINUTES

TIME TAKEN..........................

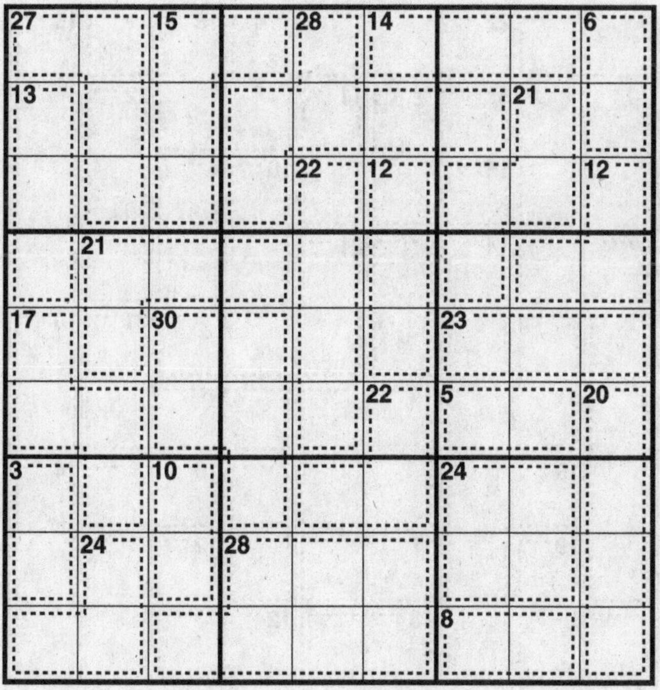

1 HOUR 20 MINUTES

TIME TAKEN..........................

Deadly

44

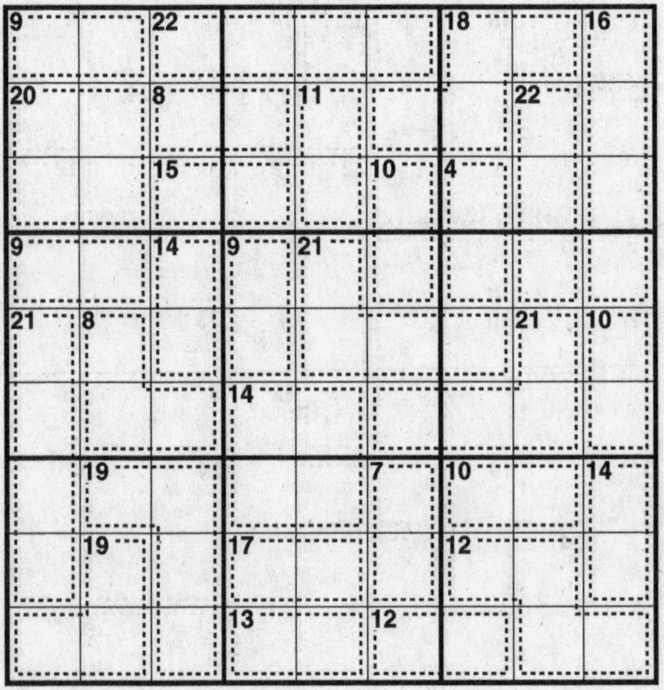

1 HOUR 20 MINUTES

TIME TAKEN........................

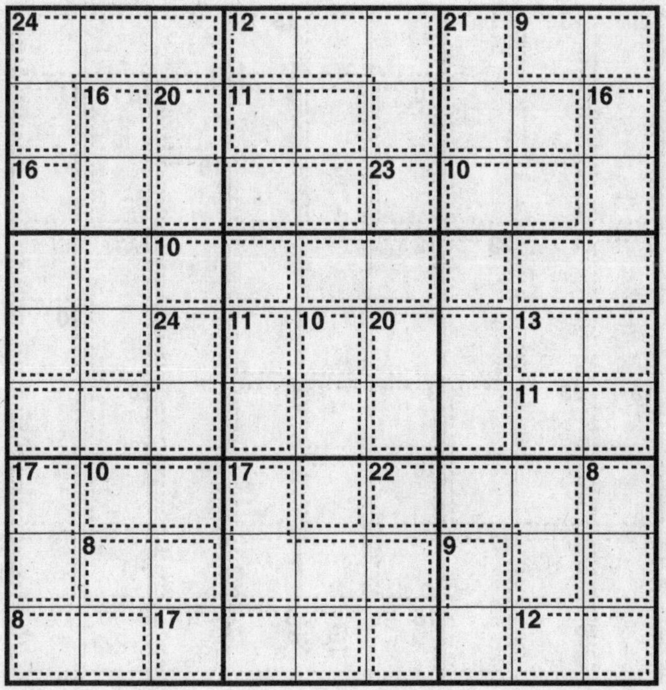

1 HOUR 20 MINUTES

TIME TAKEN..........................

Deadly

46

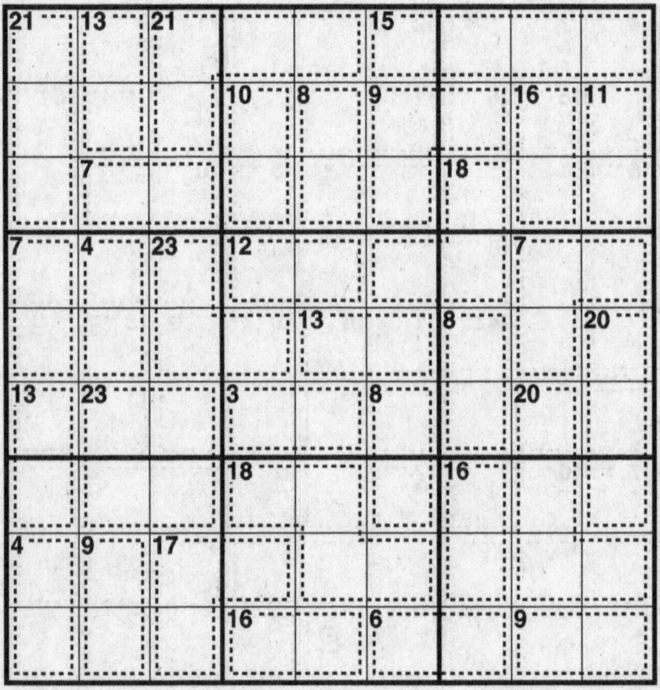

1 HOUR 20 MINUTES

TIME TAKEN..........................

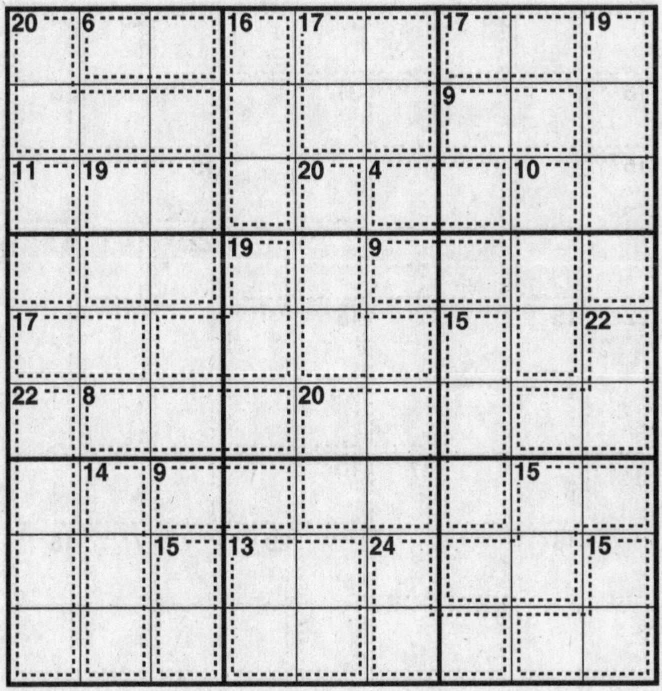

1 HOUR 20 MINUTES

TIME TAKEN............................

Deadly

48

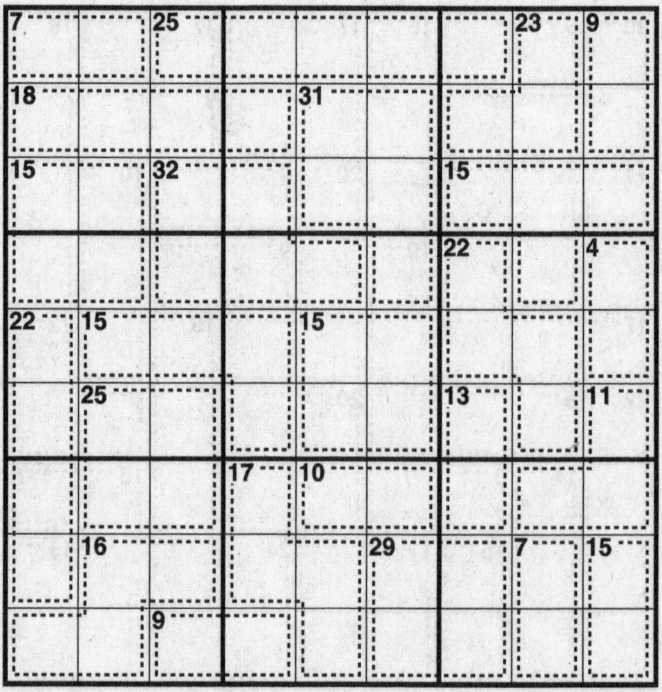

1 HOUR 20 MINUTES

TIME TAKEN..........................

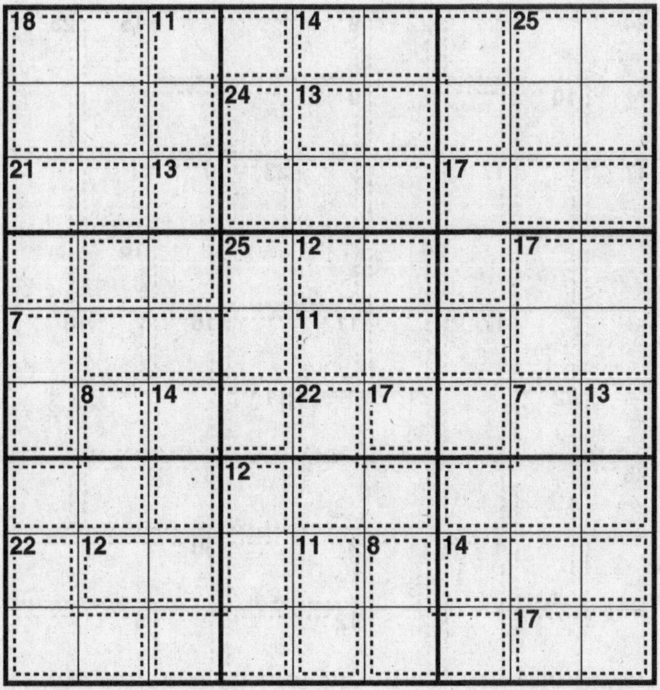

1 HOUR 20 MINUTES

TIME TAKEN..........................

Deadly

50

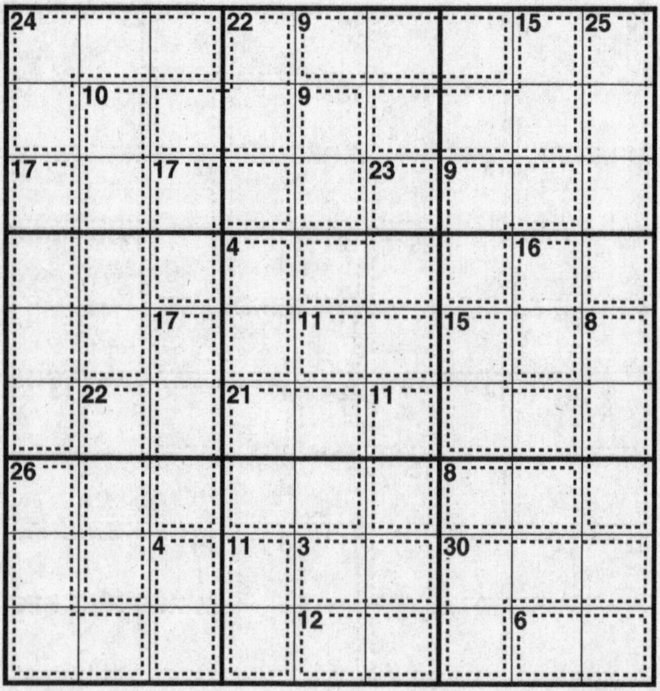

1 HOUR 20 MINUTES

TIME TAKEN..........................

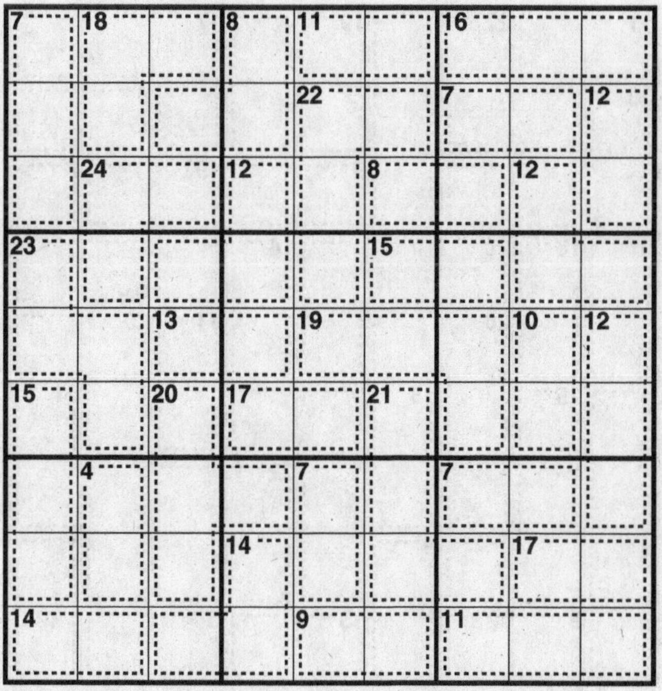

1 HOUR 20 MINUTES

TIME TAKEN..........................

Deadly

52

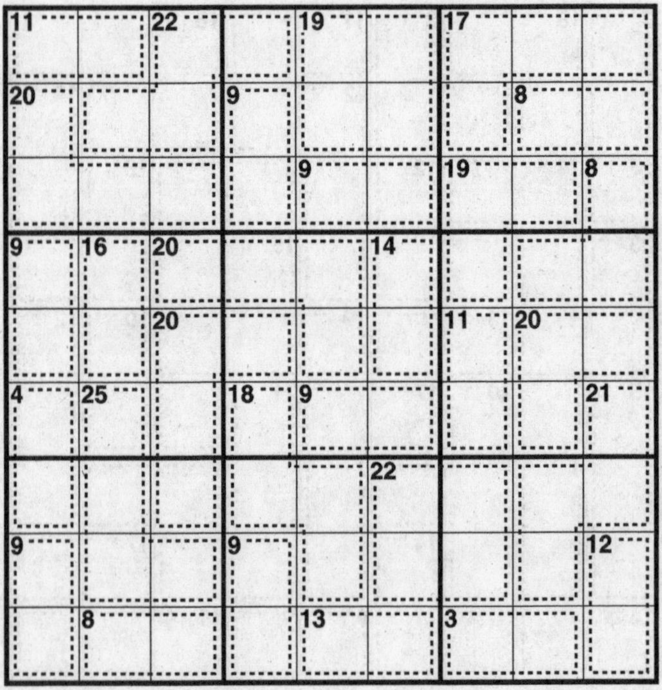

1 HOUR 25 MINUTES

TIME TAKEN.........................

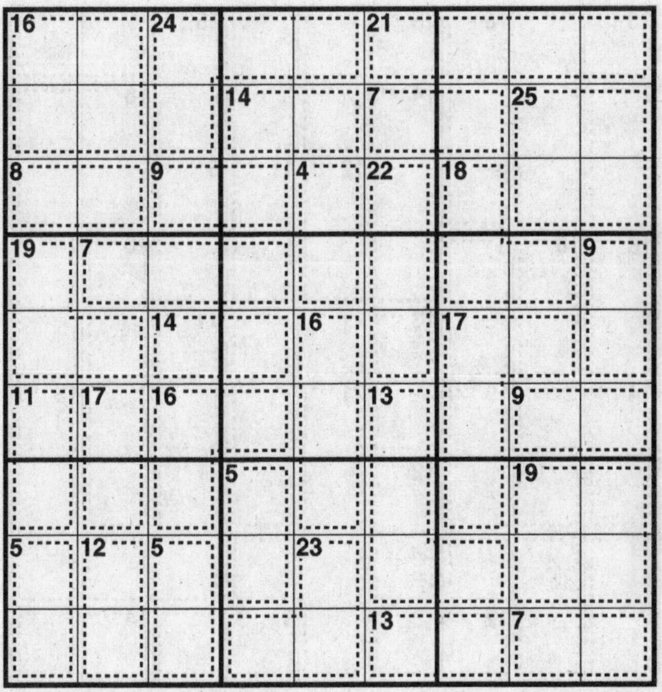

1 HOUR 25 MINUTES

TIME TAKEN..........................

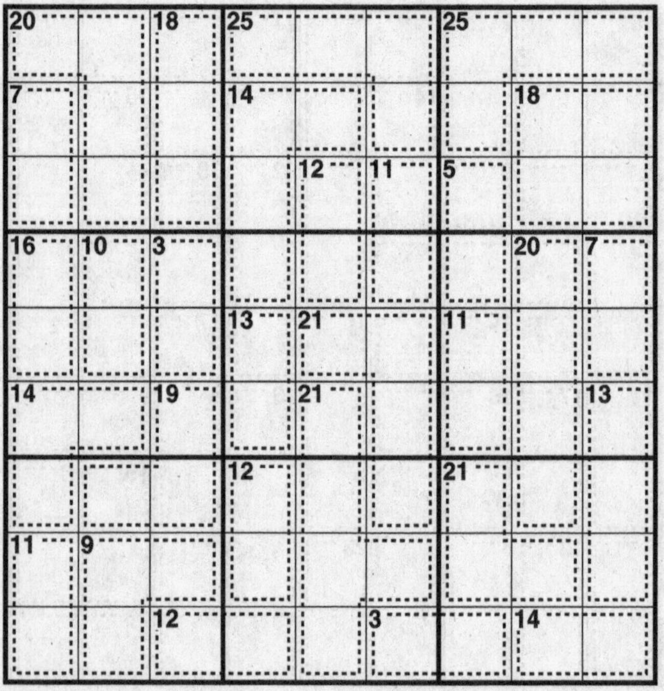

1 HOUR 25 MINUTES

TIME TAKEN.............................

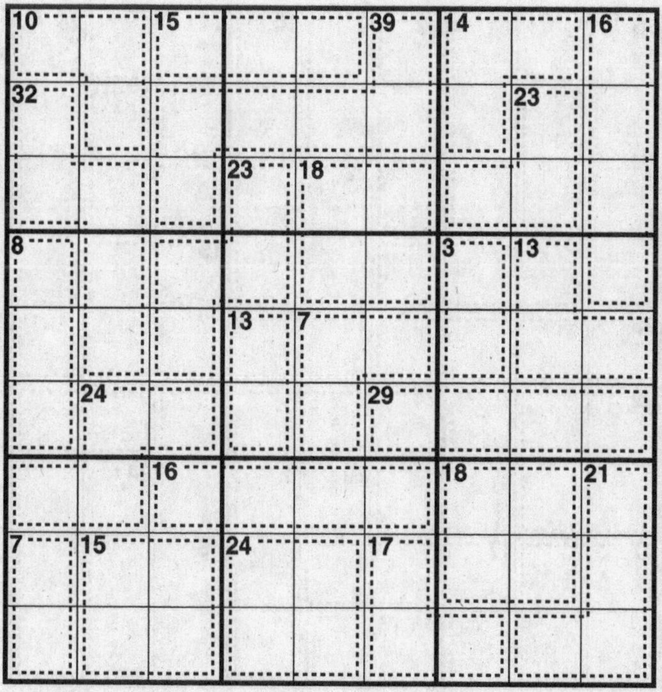

1 HOUR 25 MINUTES

TIME TAKEN.........................

Deadly

56

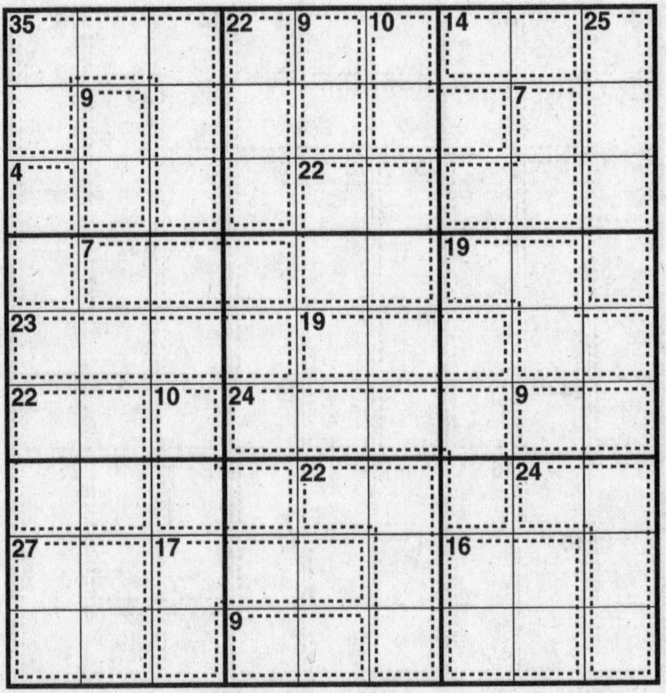

1 HOUR 25 MINUTES

TIME TAKEN..........................

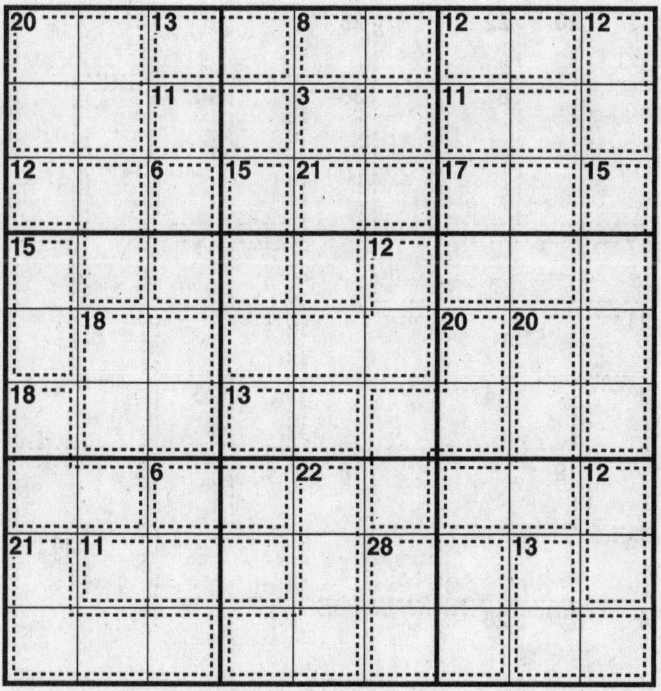

1 HOUR 25 MINUTES

TIME TAKEN..........................

Deadly

58

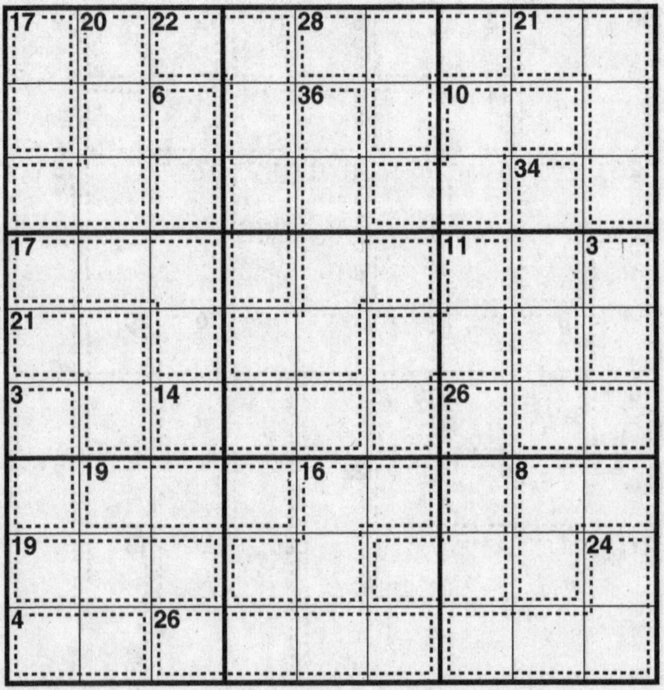

1 HOUR 25 MINUTES

TIME TAKEN..........................

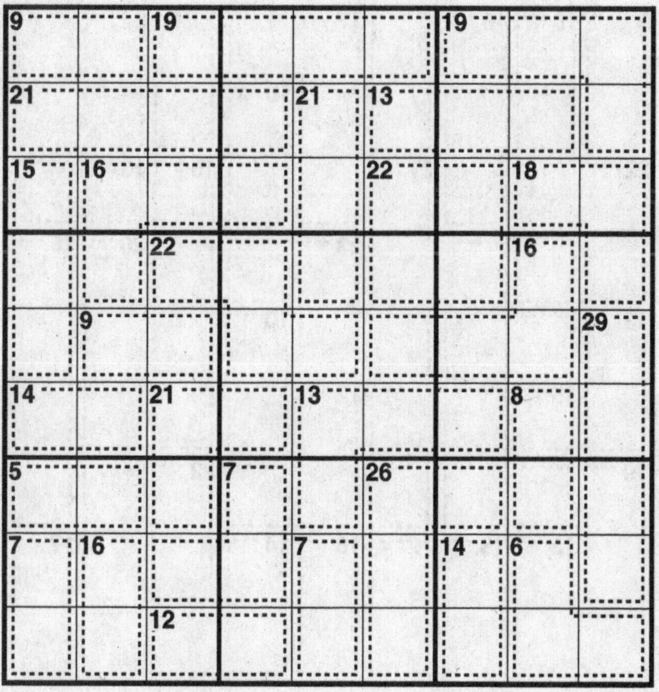

1 HOUR 25 MINUTES

TIME TAKEN............................

60

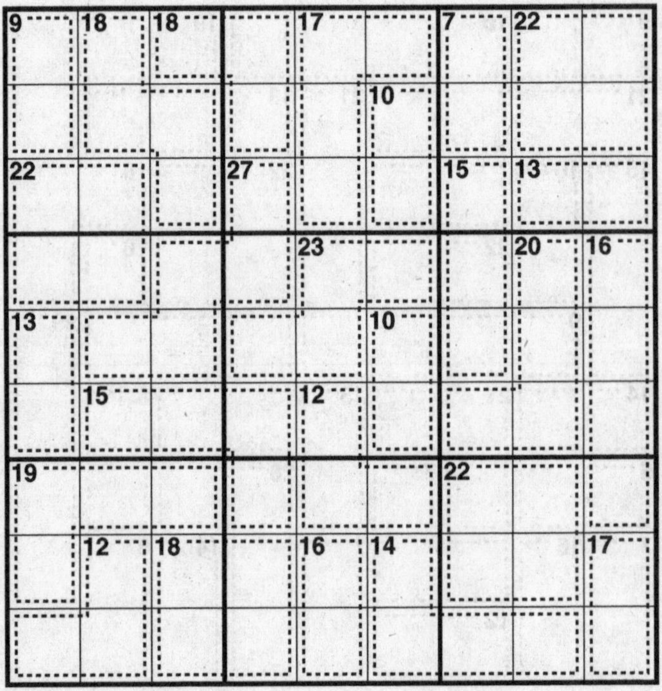

1 HOUR 25 MINUTES

TIME TAKEN..........................

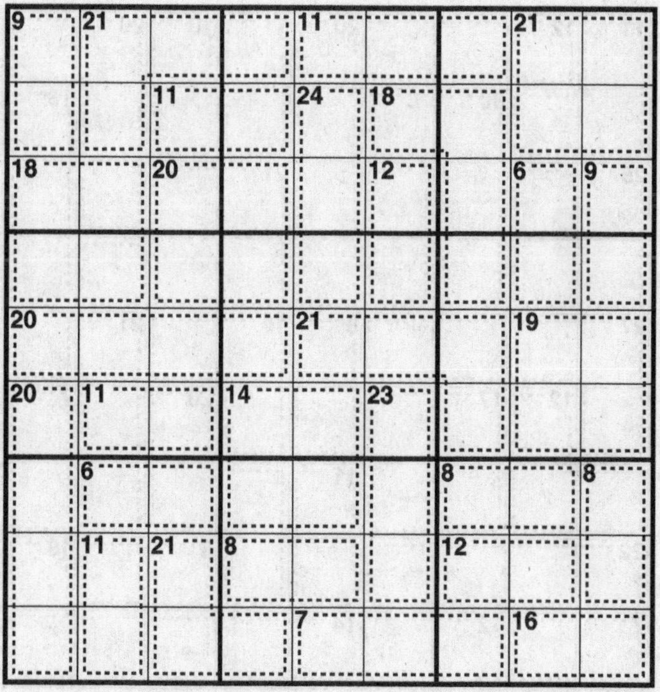

1 HOUR 25 MINUTES

TIME TAKEN..............................

62

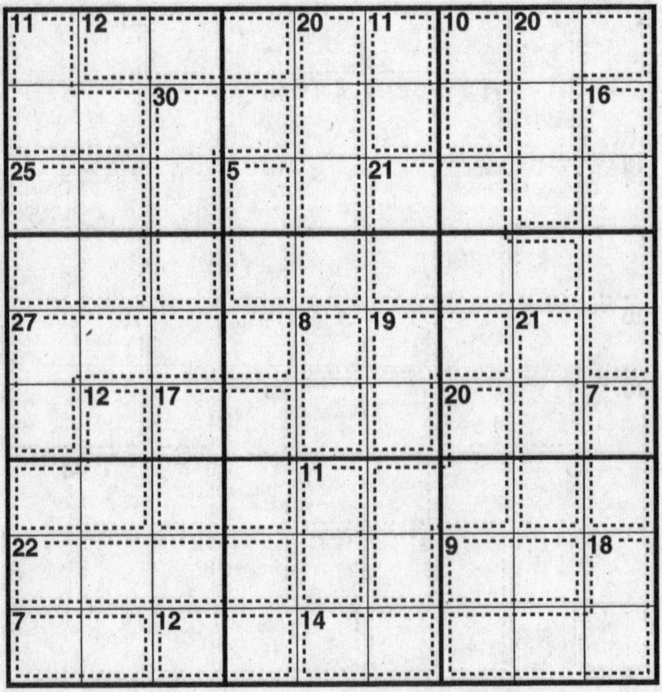

1 HOUR 25 MINUTES

TIME TAKEN.........................

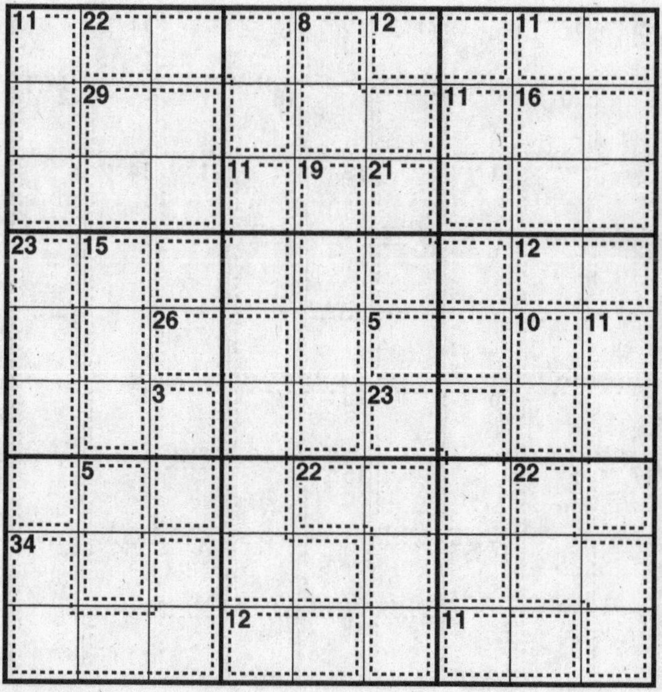

1 HOUR 25 MINUTES

TIME TAKEN..............................

Deadly

64

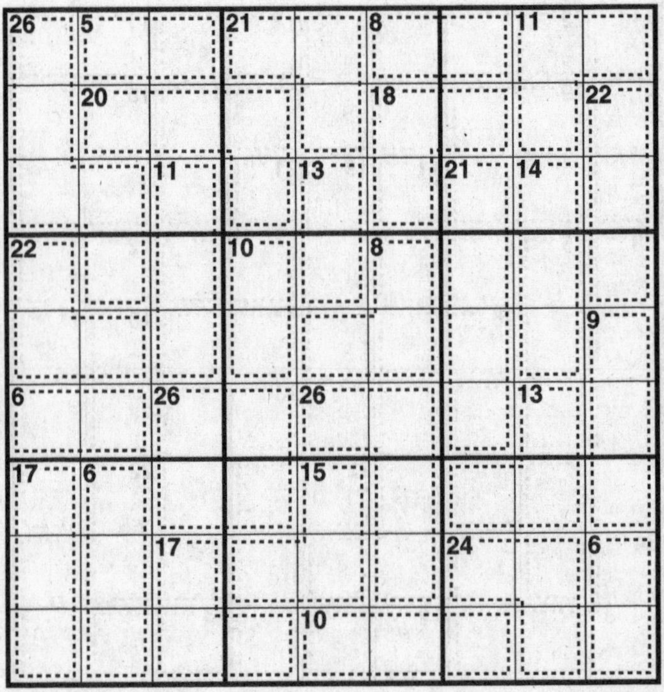

1 HOUR 30 MINUTES

TIME TAKEN..........................

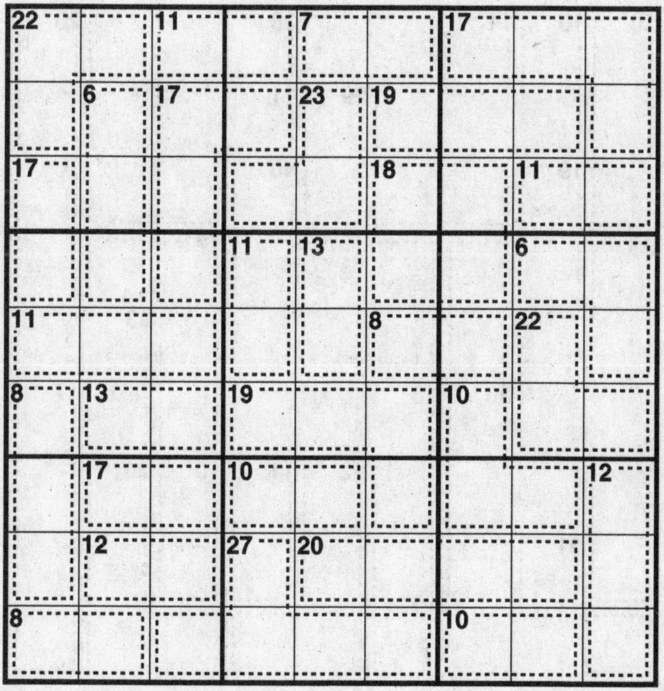

1 HOUR 30 MINUTES

TIME TAKEN..........................

66

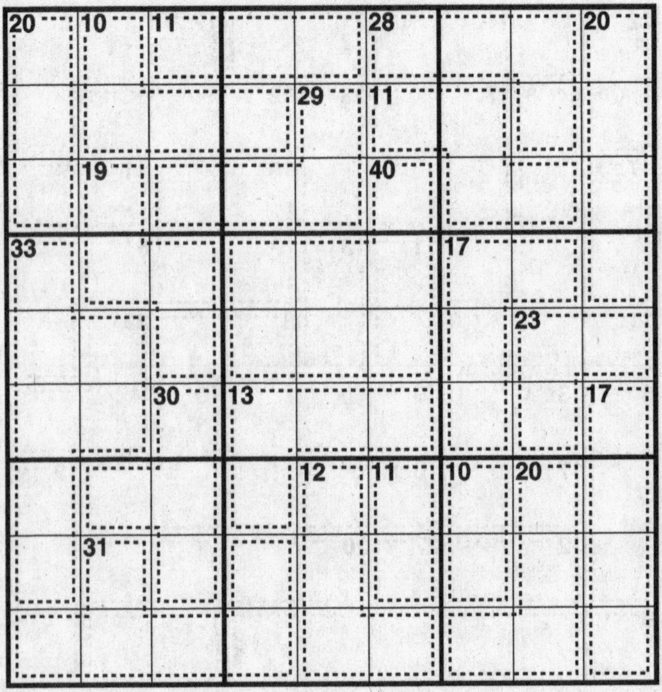

1 HOUR 30 MINUTES

TIME TAKEN..........................

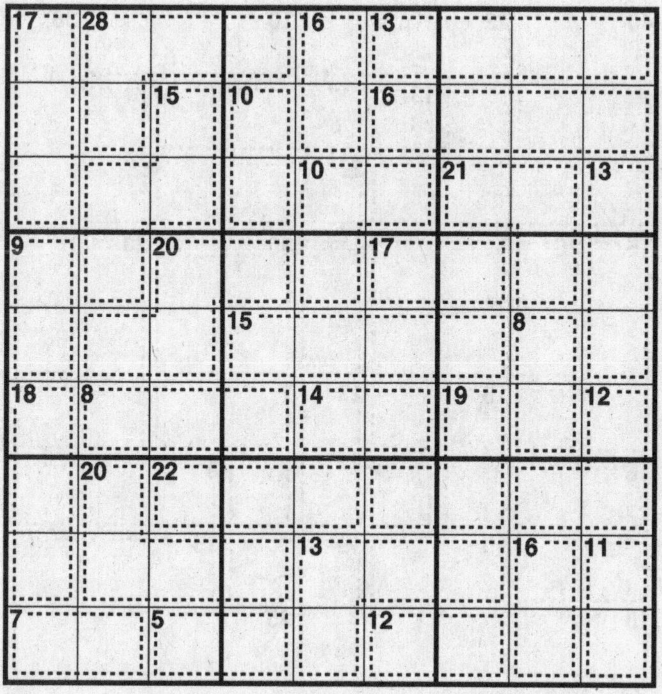

1 HOUR 30 MINUTES

TIME TAKEN.........................

Deadly

68

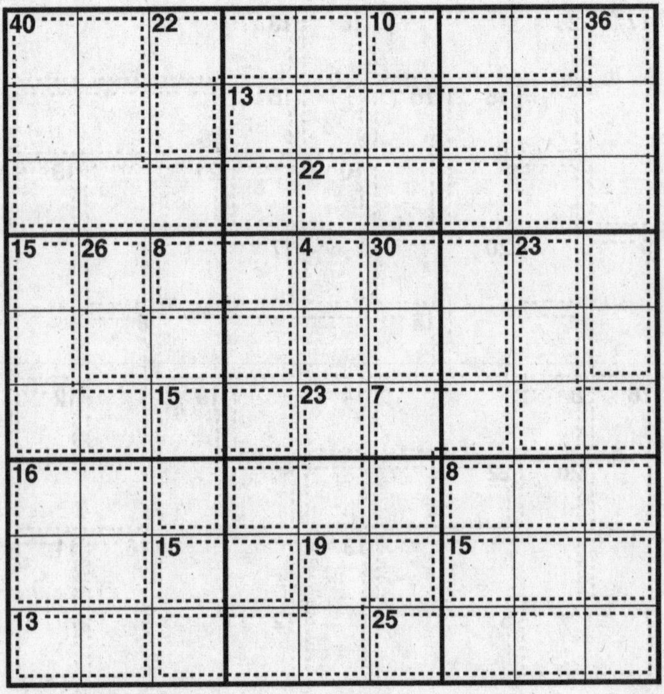

1 HOUR 30 MINUTES

TIME TAKEN..........................

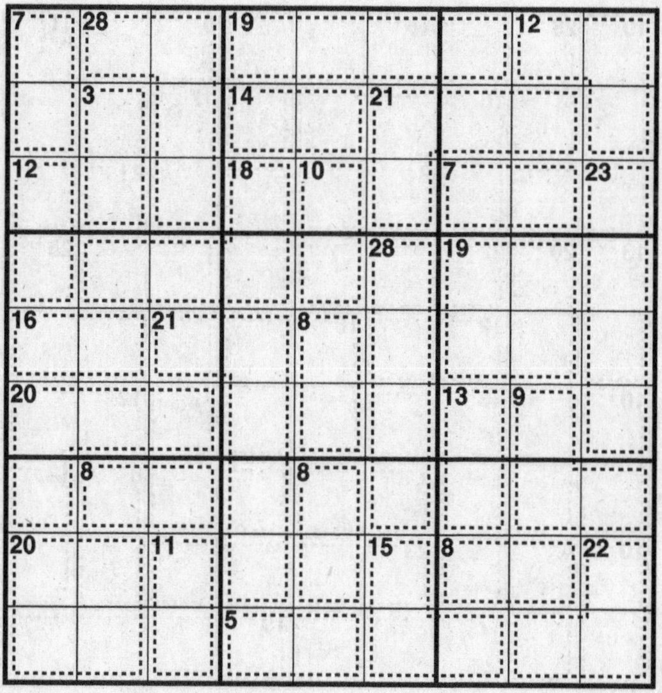

1 HOUR 30 MINUTES

TIME TAKEN..........................

Deadly

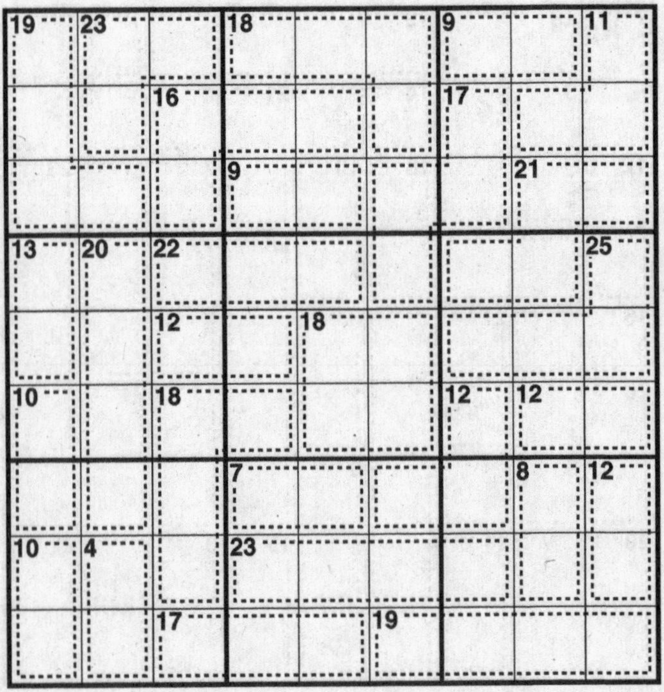

1 HOUR 30 MINUTES

TIME TAKEN..........................

71

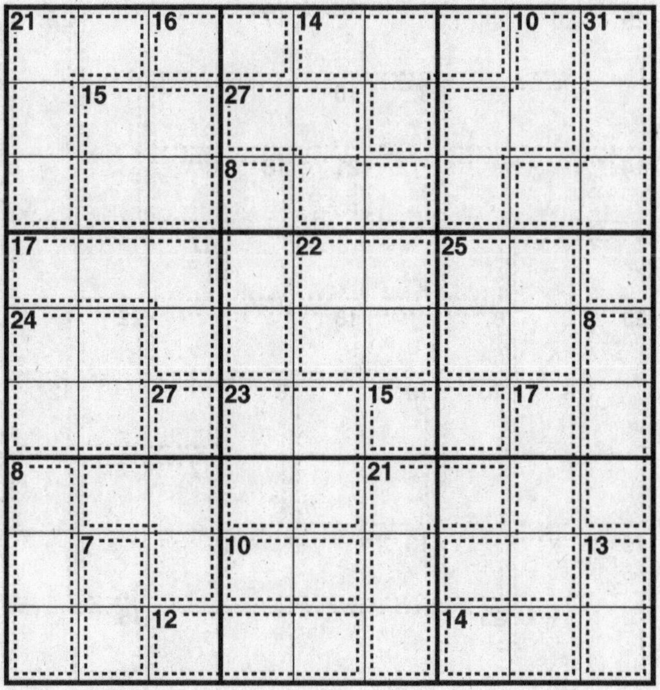

1 HOUR 30 MINUTES

TIME TAKEN............................

Deadly

72

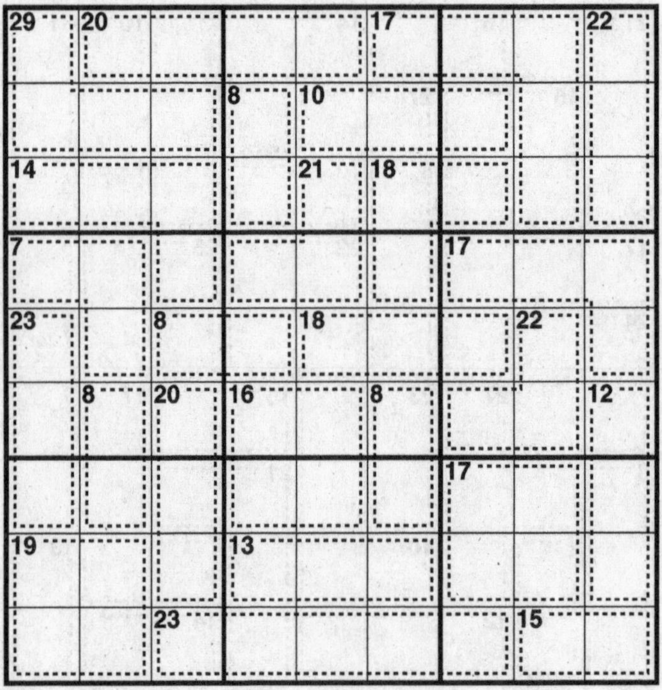

1 HOUR 30 MINUTES

TIME TAKEN............................

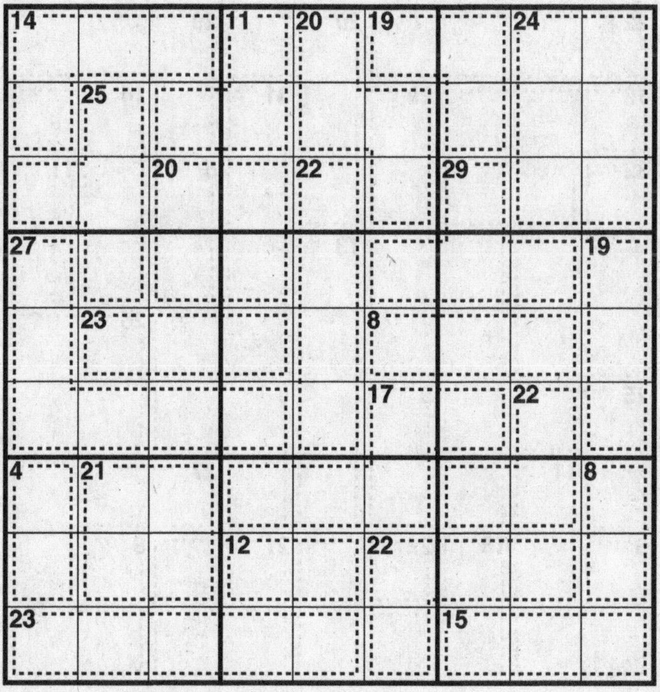

1 HOUR 30 MINUTES

TIME TAKEN..............................

74

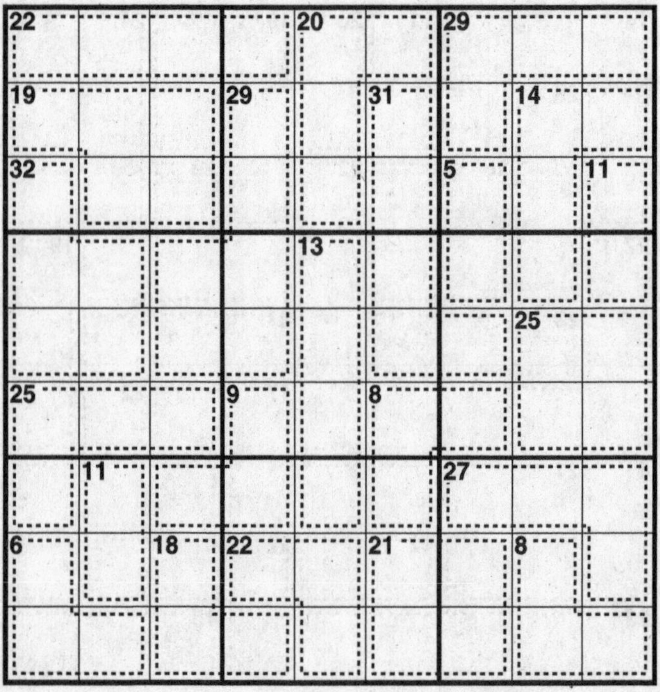

1 HOUR 30 MINUTES

TIME TAKEN............................

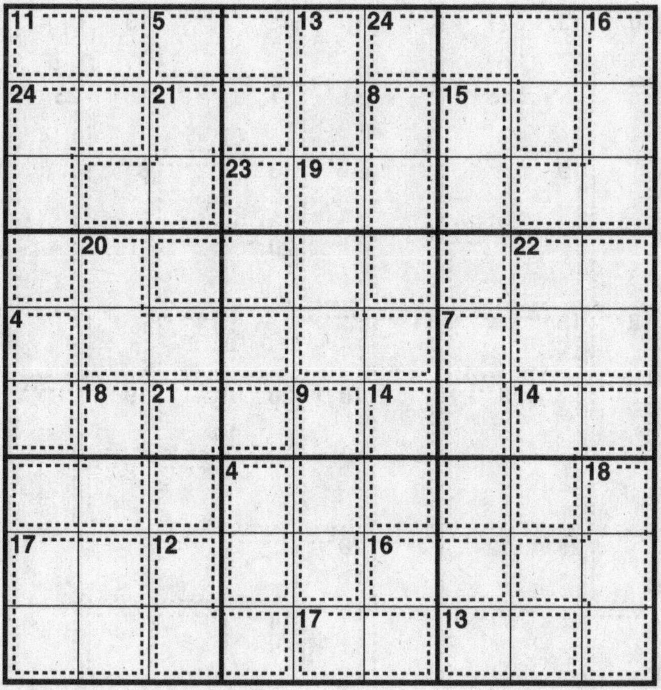

1 HOUR 30 MINUTES

TIME TAKEN.........................

76

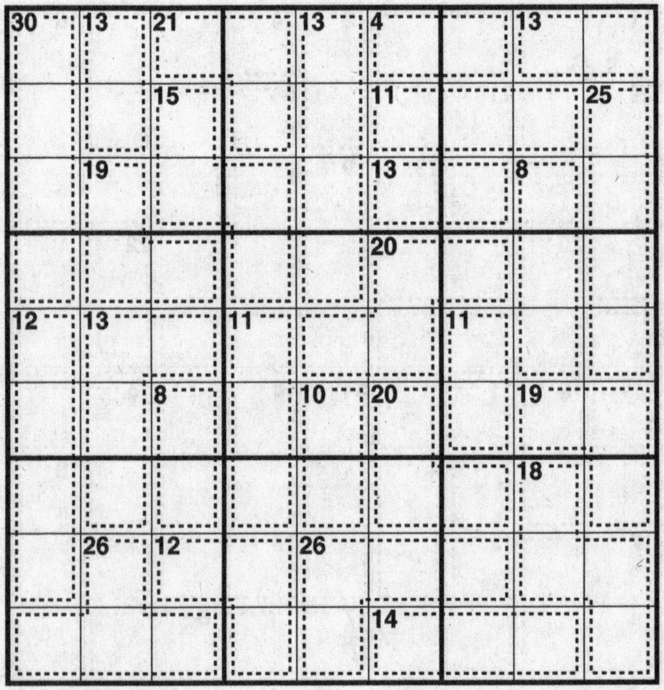

1 HOUR 30 MINUTES

TIME TAKEN..........................

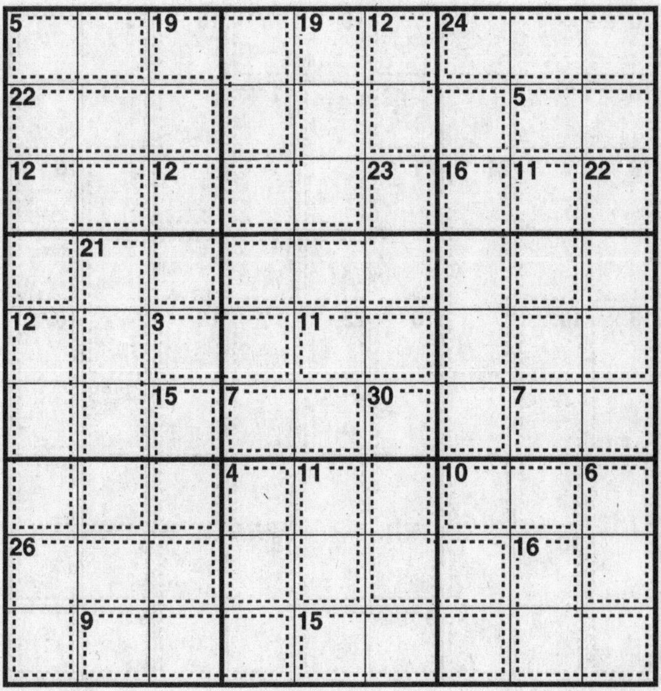

1 HOUR 30 MINUTES

TIME TAKEN..........................

Deadly

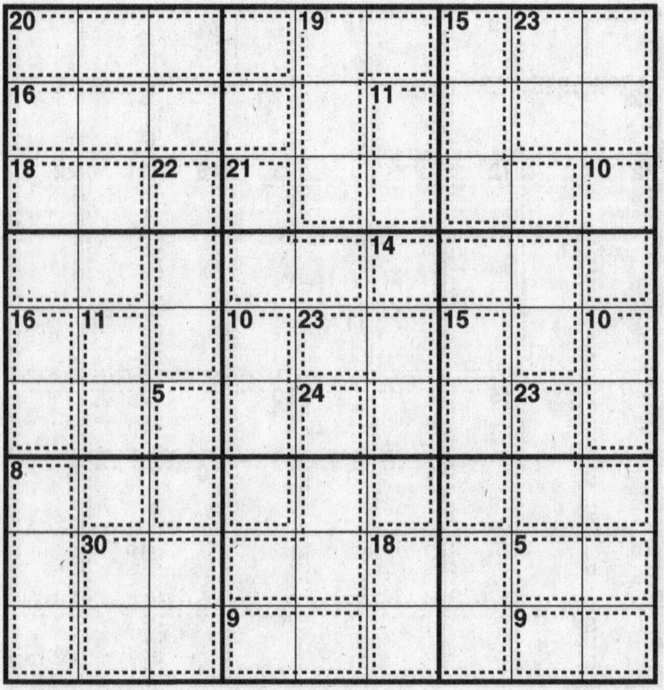

1 HOUR 30 MINUTES

TIME TAKEN............................

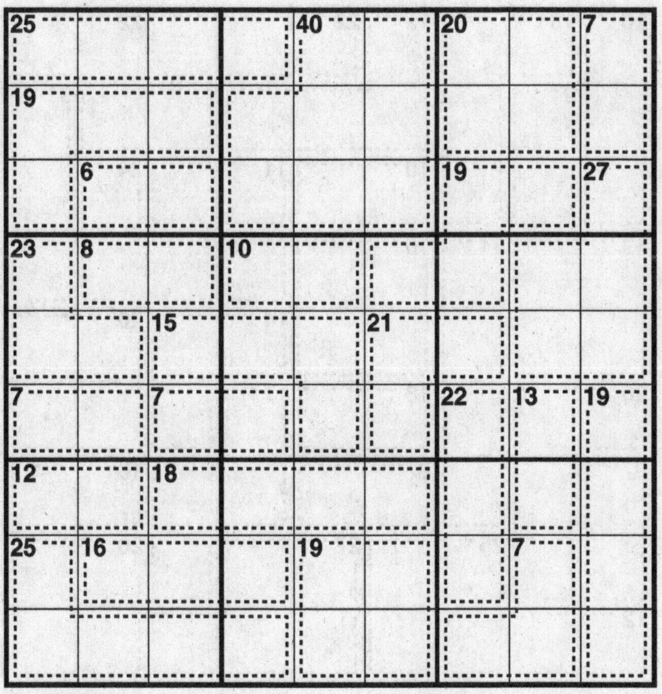

1 HOUR 30 MINUTES

TIME TAKEN..........................

Deadly

80

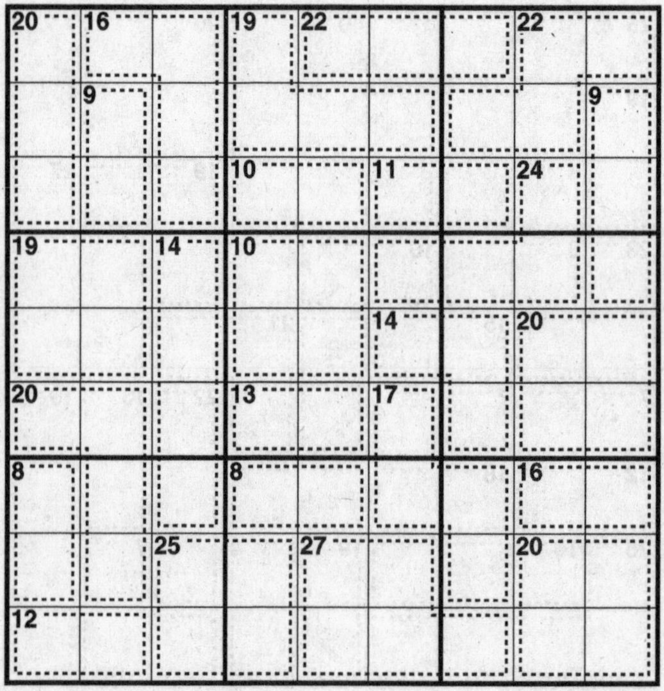

1 HOUR 30 MINUTES

TIME TAKEN..............................

81

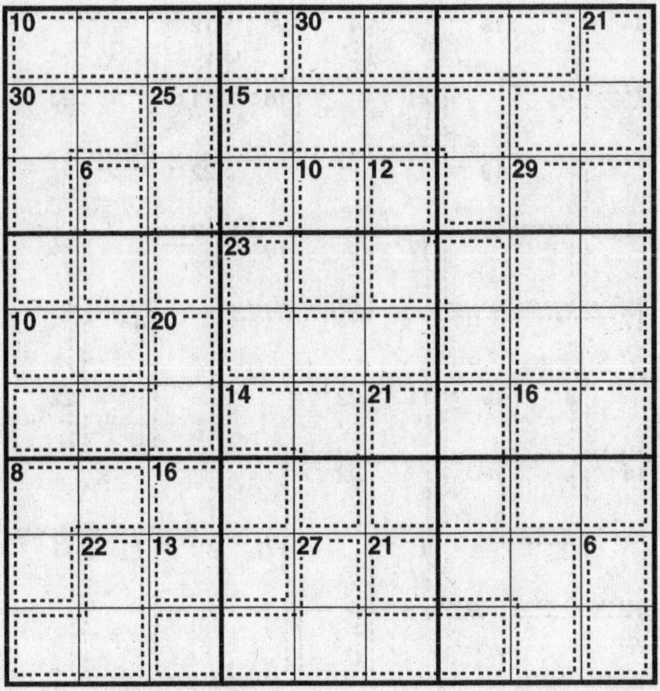

1 HOUR 30 MINUTES

TIME TAKEN............................

Deadly

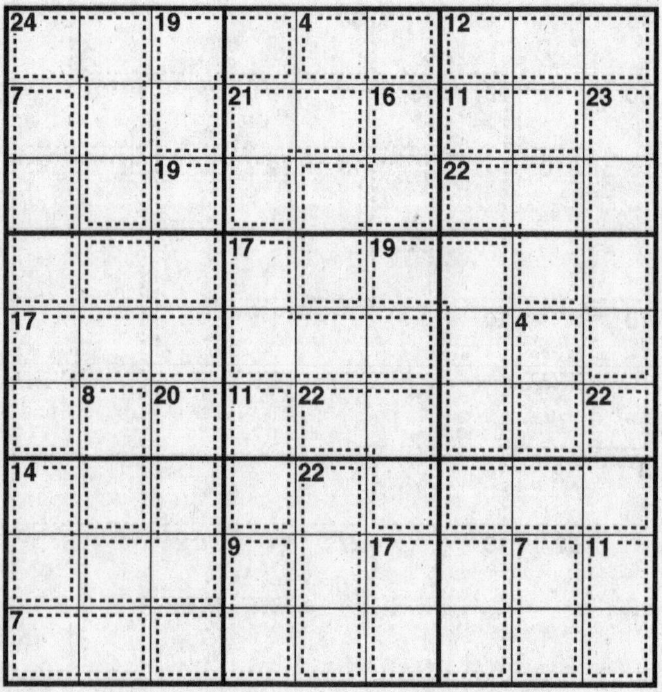

1 HOUR 30 MINUTES

TIME TAKEN..........................

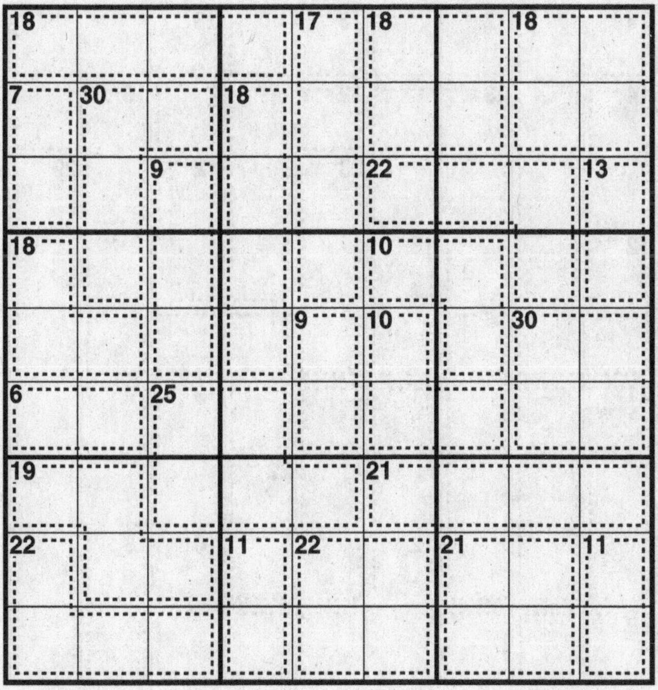

1 HOUR 30 MINUTES

TIME TAKEN..........................

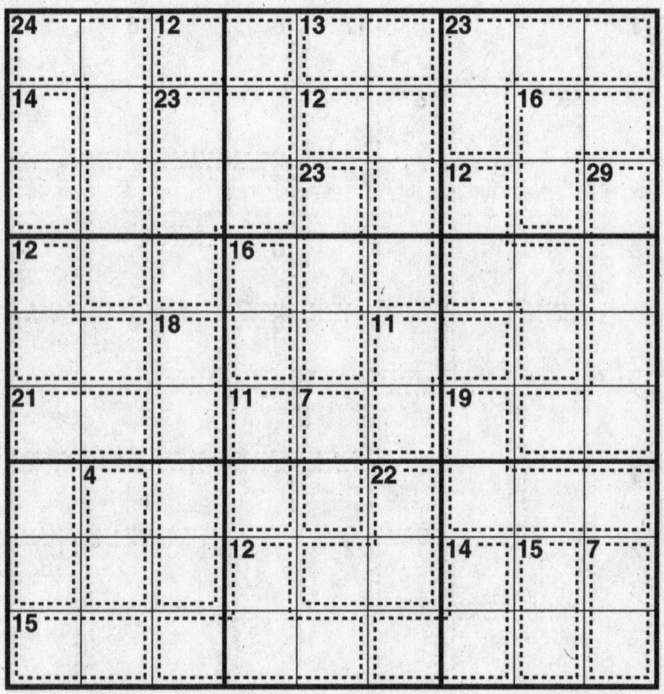

1 HOUR 35 MINUTES

TIME TAKEN..........................

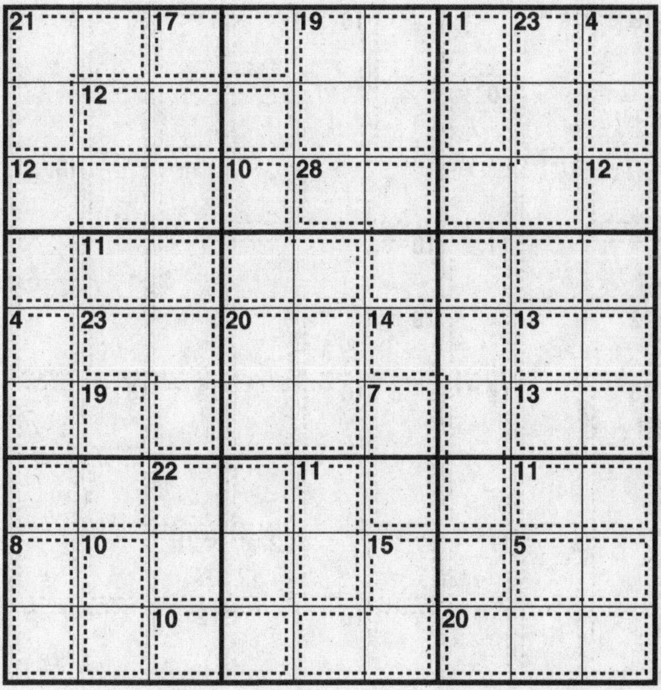

1 HOUR 35 MINUTES

TIME TAKEN............................

86

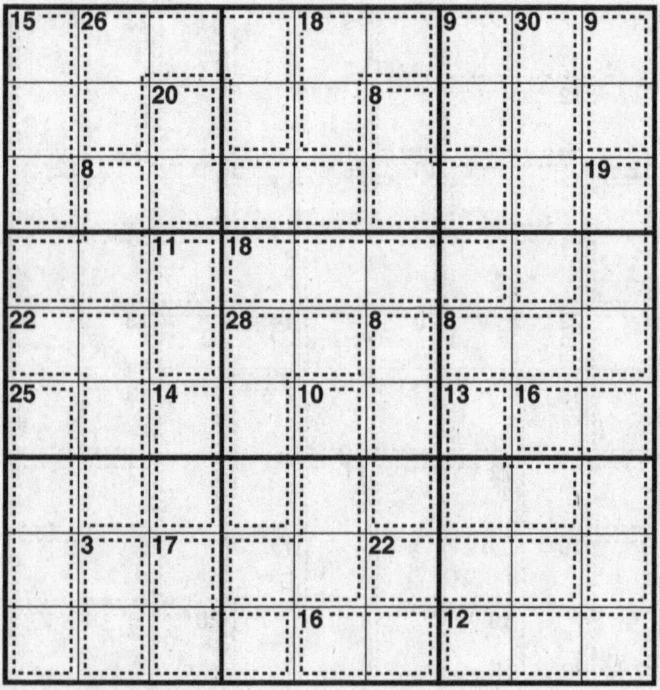

1 HOUR 35 MINUTES

TIME TAKEN..........................

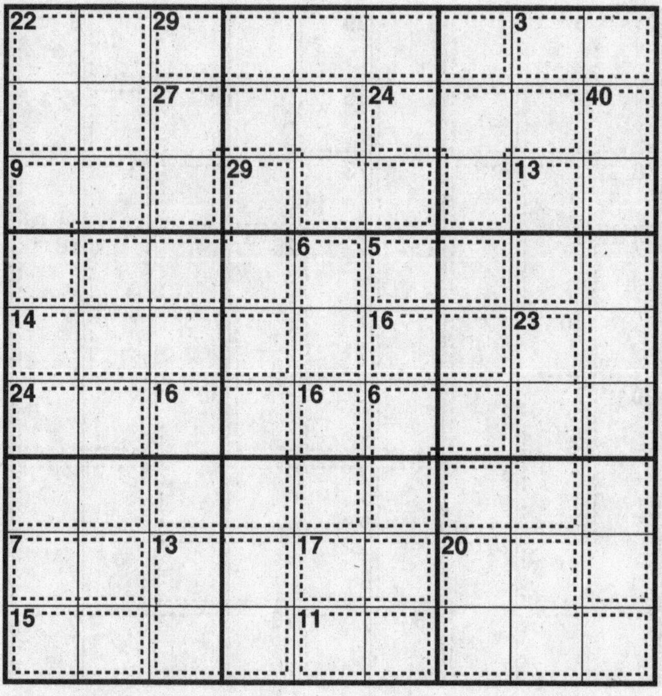

1 HOUR 35 MINUTES

TIME TAKEN..........................

88

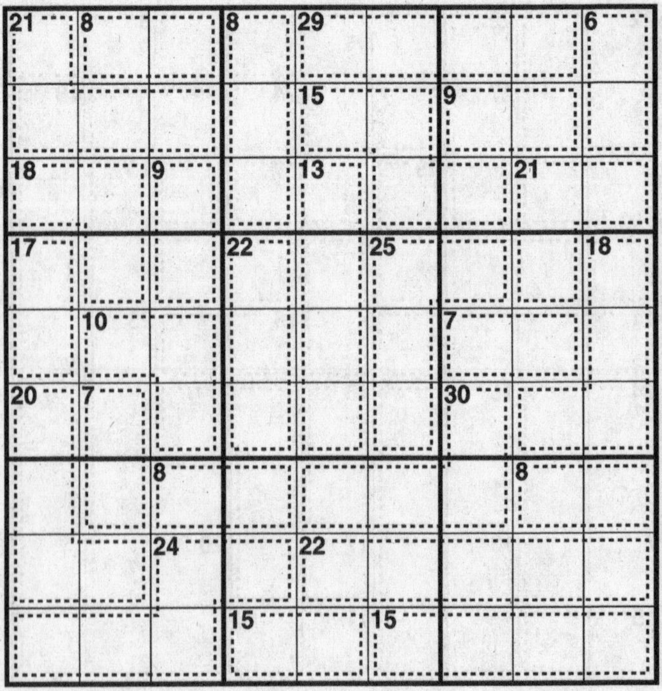

1 HOUR 35 MINUTES

TIME TAKEN..........................

89

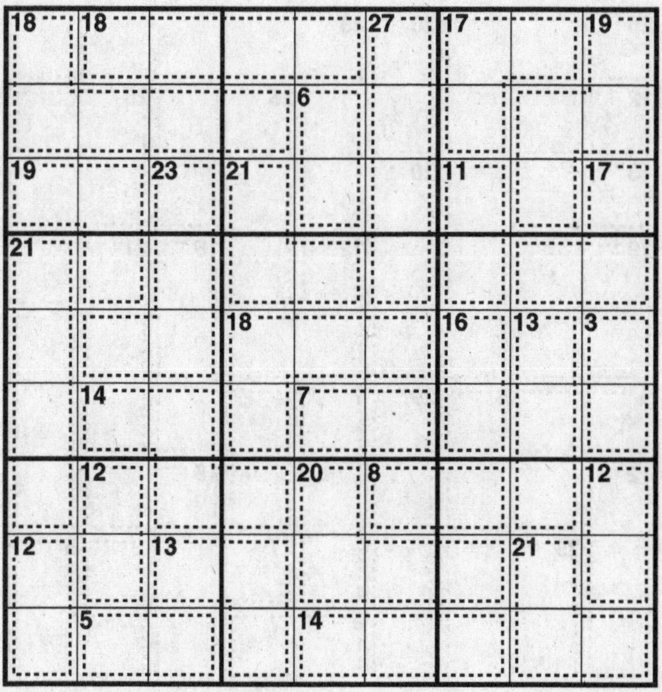

1 HOUR 35 MINUTES

TIME TAKEN............................

Deadly

90

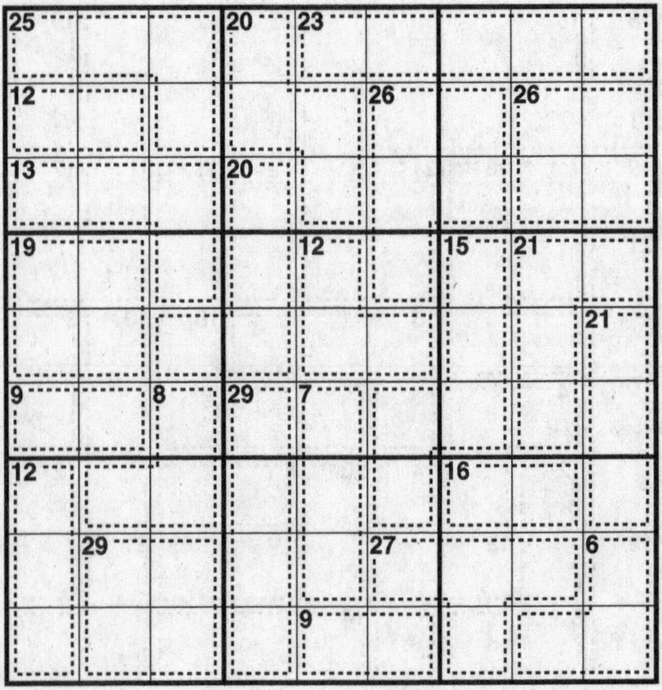

1 HOUR 40 MINUTES

TIME TAKEN..........................

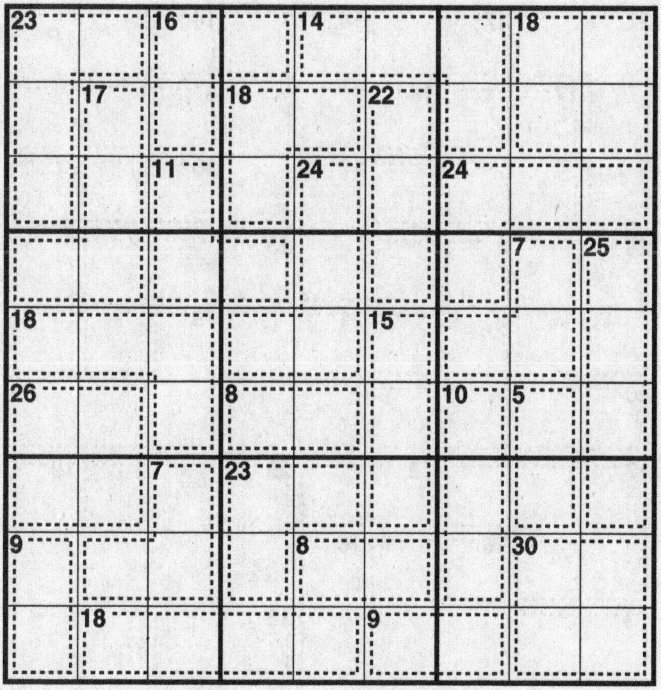

1 HOUR 40 MINUTES

TIME TAKEN..........................

Deadly

92

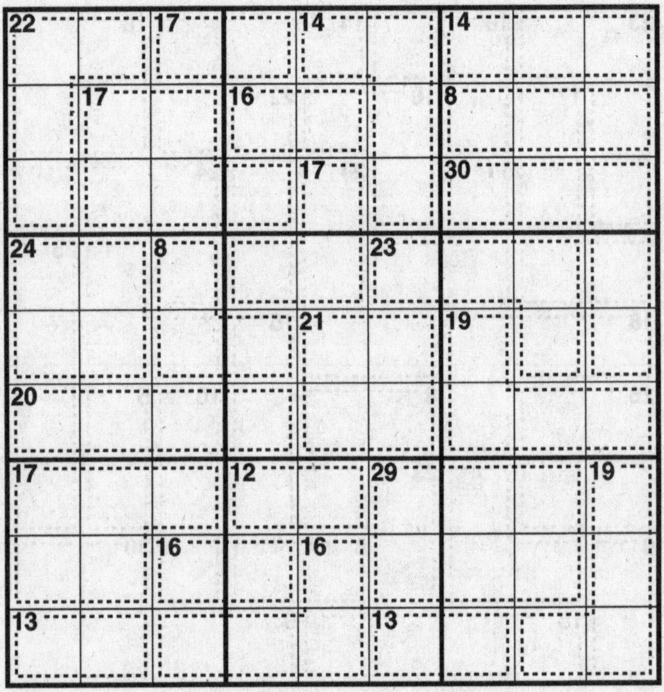

1 HOUR 40 MINUTES

TIME TAKEN.........................

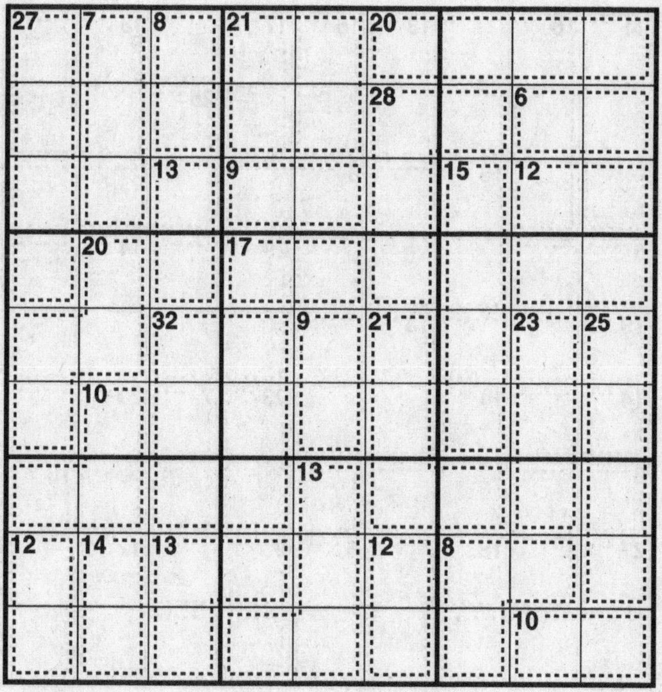

1 HOUR 40 MINUTES

TIME TAKEN............................

Deadly

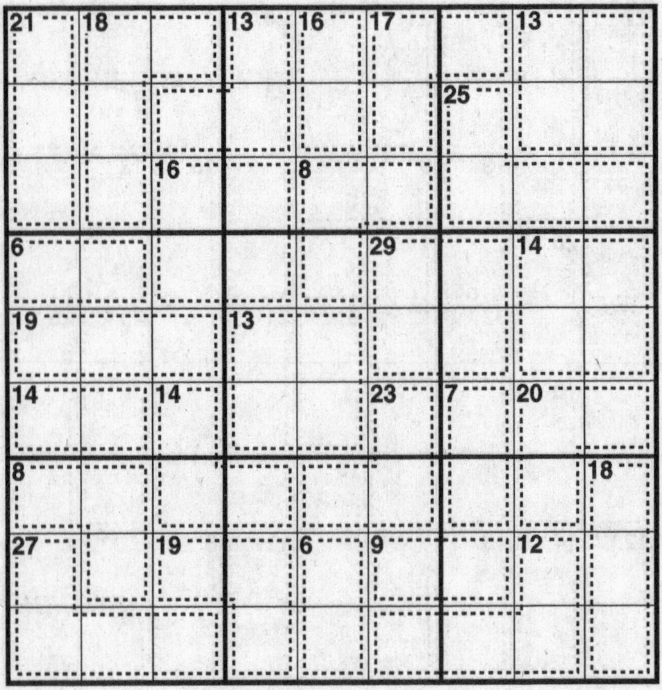

1 HOUR 50 MINUTES

TIME TAKEN..........................

95

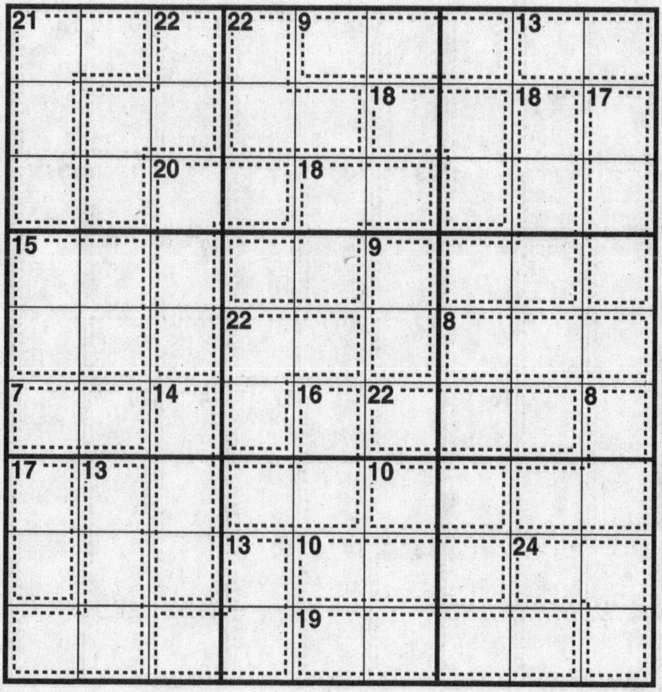

1 HOUR 50 MINUTES

TIME TAKEN..............................

Deadly

96

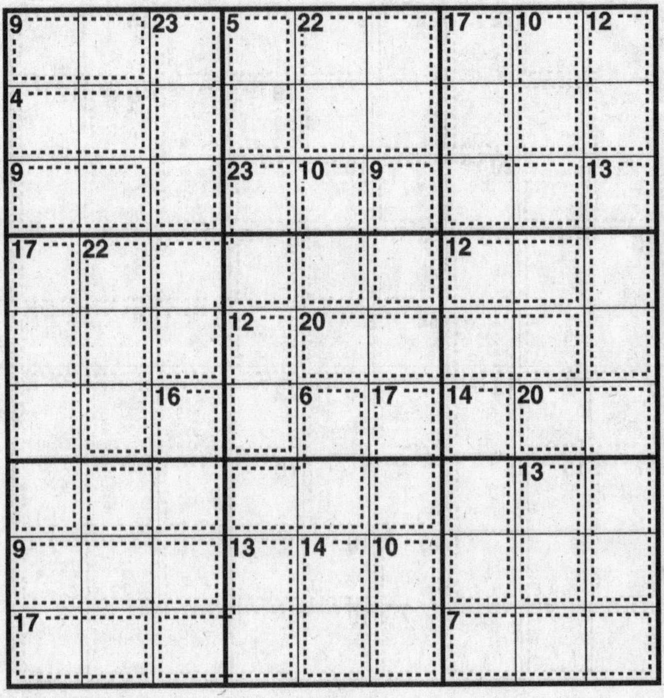

1 HOUR 50 MINUTES

TIME TAKEN..........................

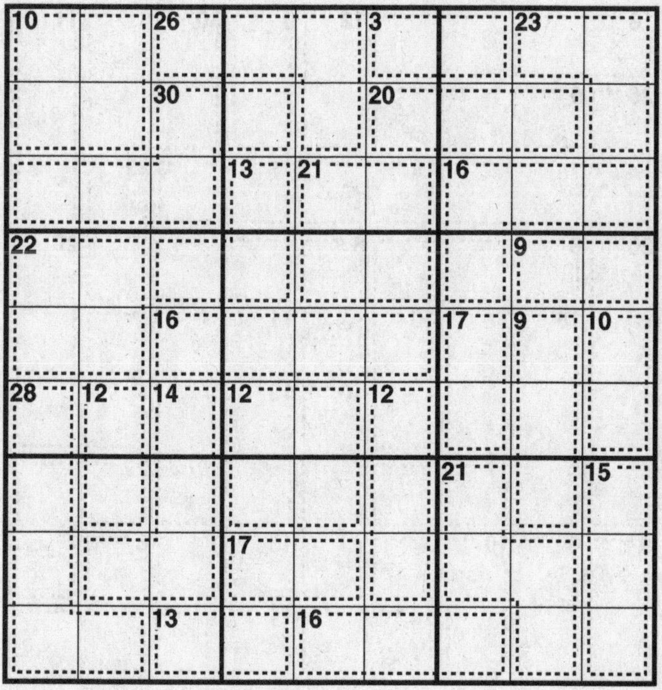

1 HOUR 50 MINUTES

TIME TAKEN..........................

Deadly

98

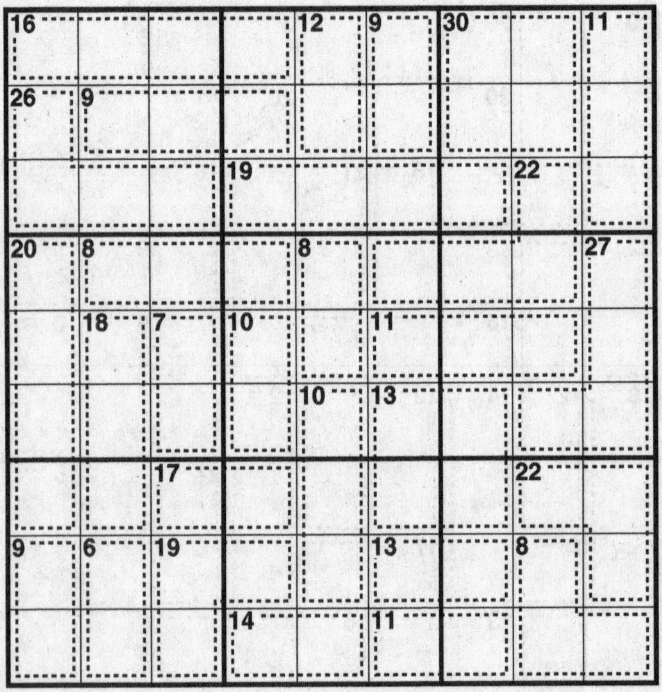

1 HOUR 50 MINUTES

TIME TAKEN.........................

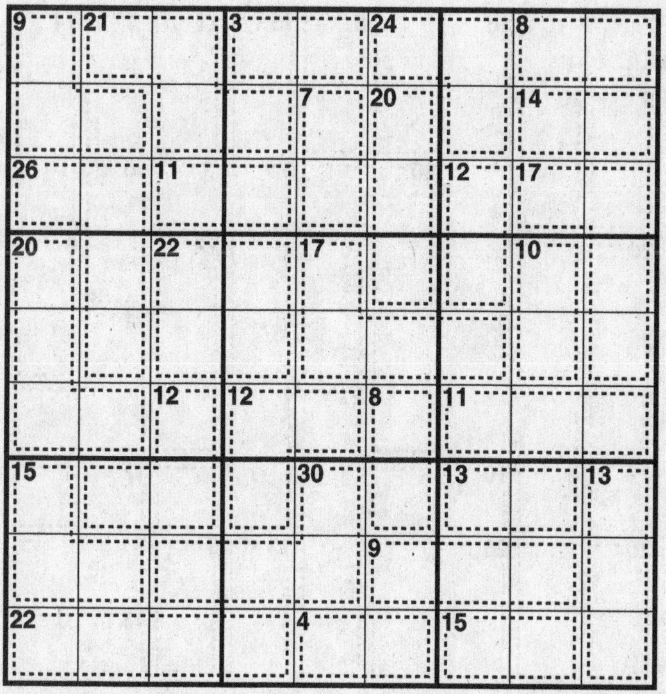

1 HOUR 50 MINUTES

TIME TAKEN..........................

Deadly

100

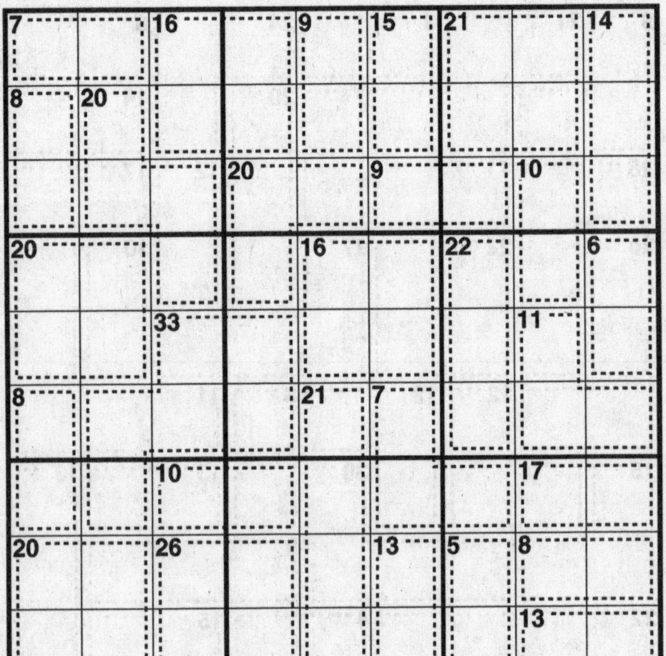

1 HOUR 50 MINUTES

TIME TAKEN.........................

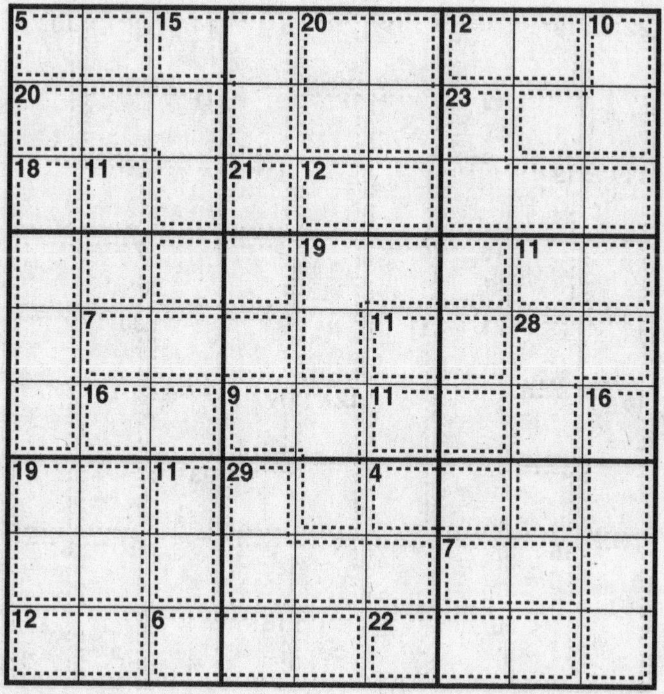

1 HOUR 55 MINUTES

TIME TAKEN..........................

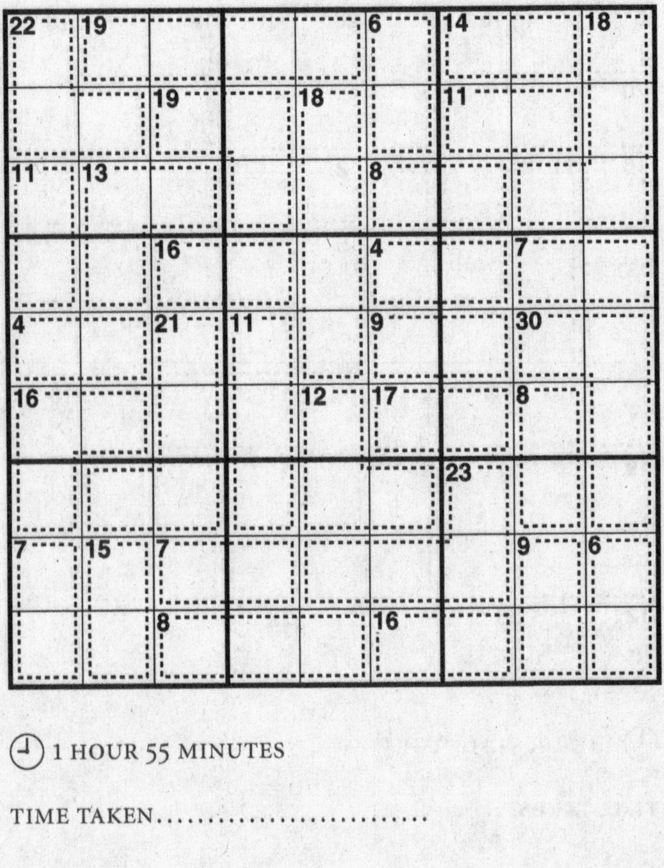

1 HOUR 55 MINUTES

TIME TAKEN.........................

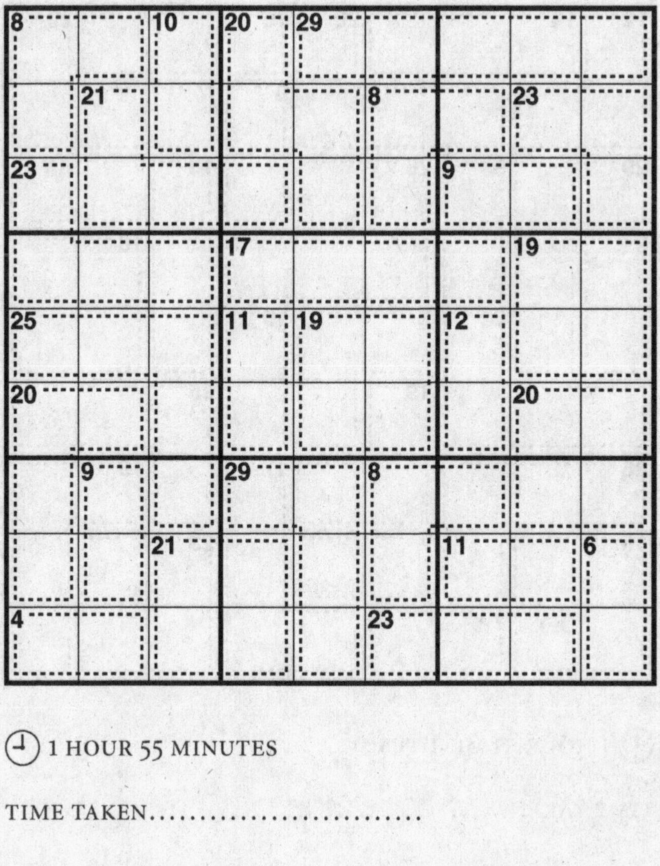

1 HOUR 55 MINUTES

TIME TAKEN..........................

104

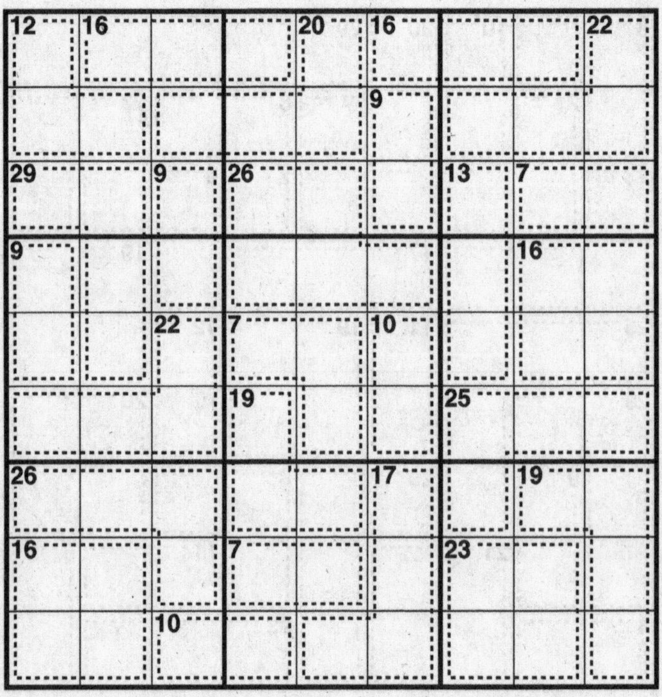

1 HOUR 55 MINUTES

TIME TAKEN.........................

105

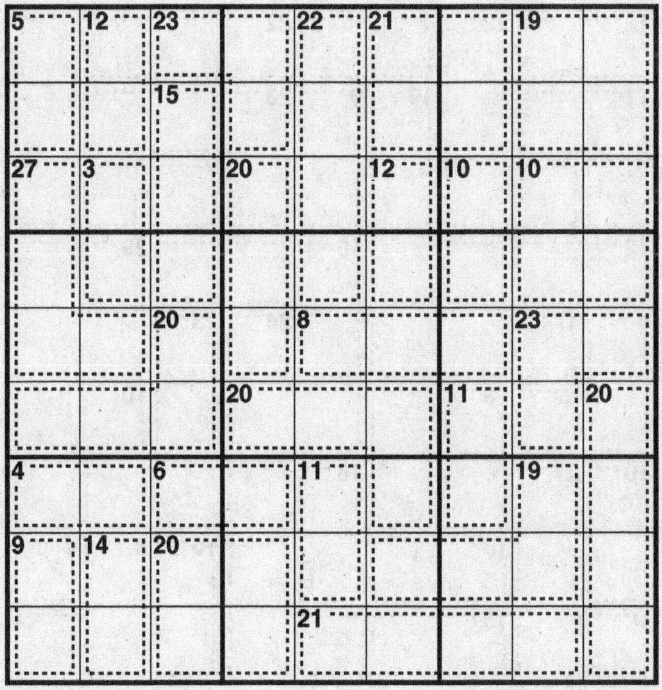

1 HOUR 55 MINUTES

TIME TAKEN..........................

Extra Deadly

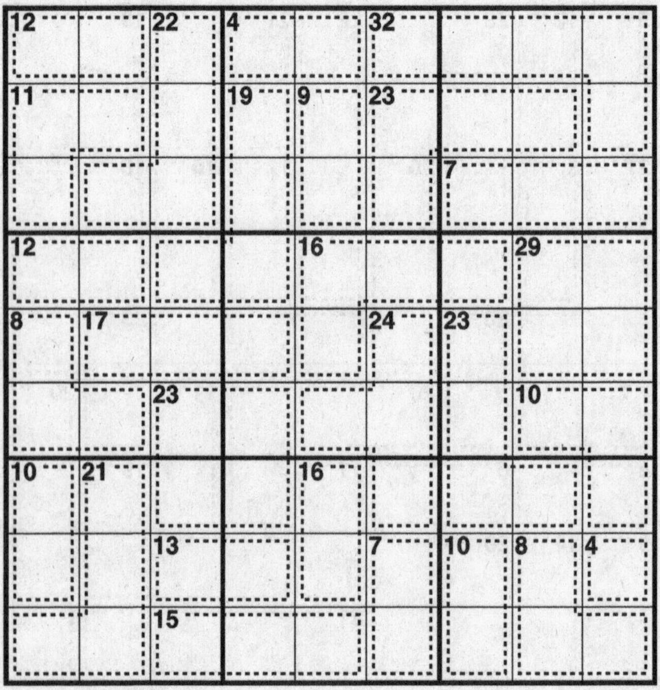

2 HOURS

TIME TAKEN.........................

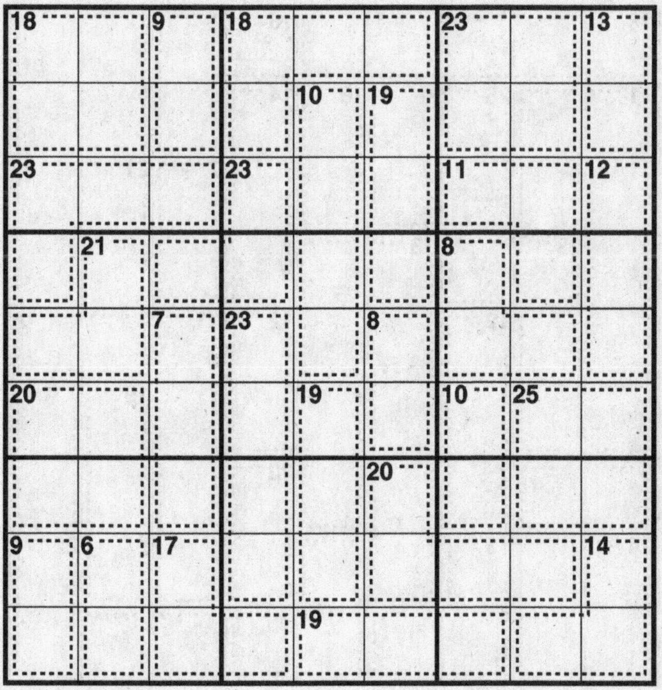

2 HOURS

TIME TAKEN............................

108

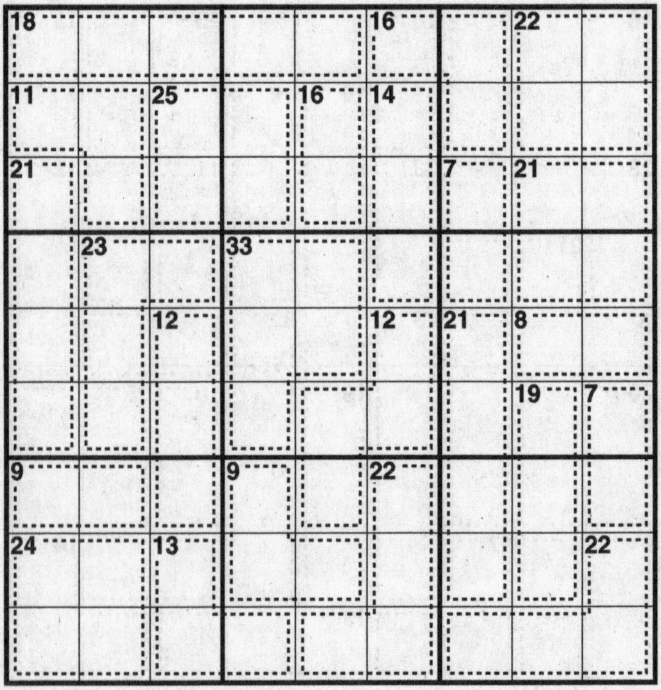

2 HOURS

TIME TAKEN..........................

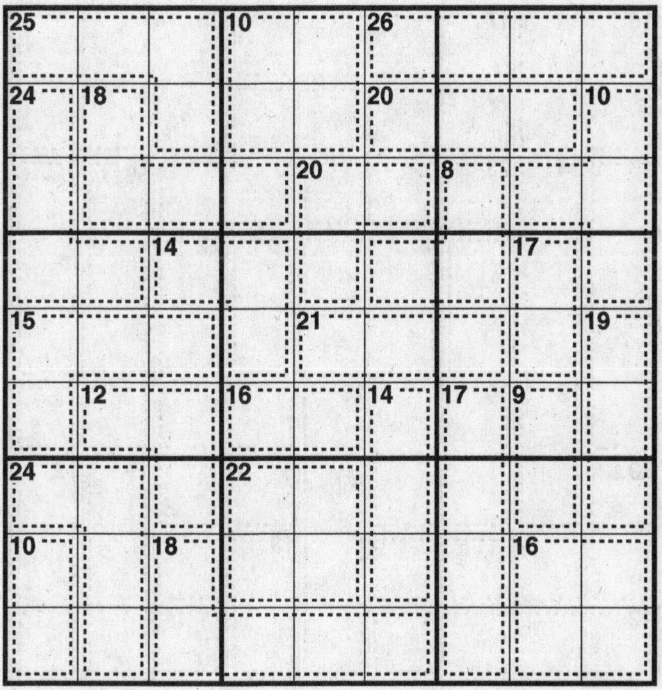

2 HOURS

TIME TAKEN.............................

110

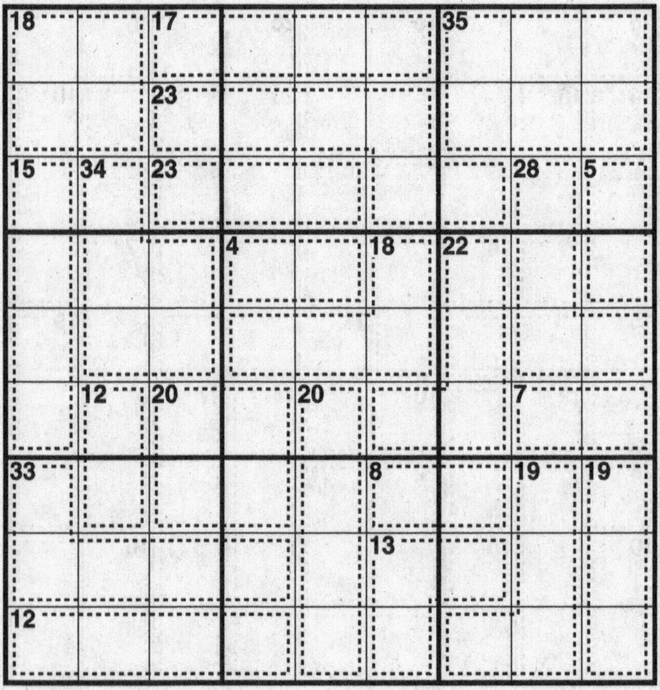

2 HOURS 15 MINUTES

TIME TAKEN............................

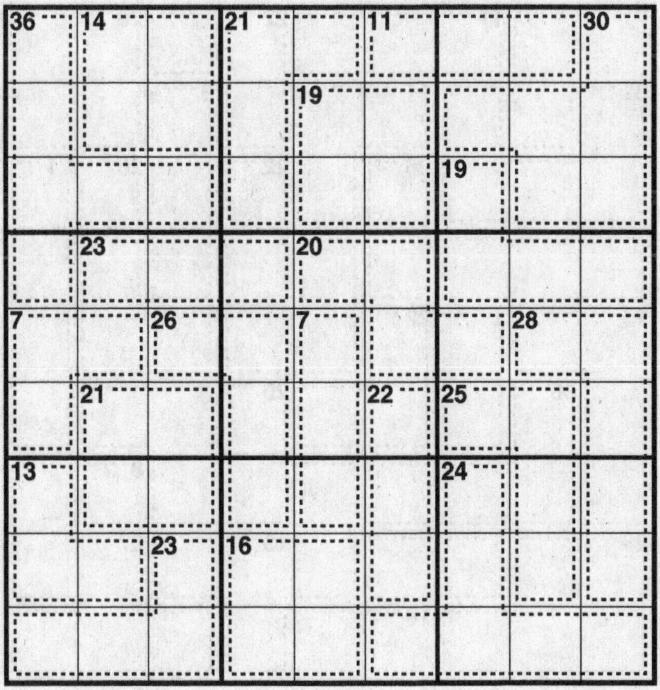

2 HOURS 15 MINUTES

TIME TAKEN.........................

112

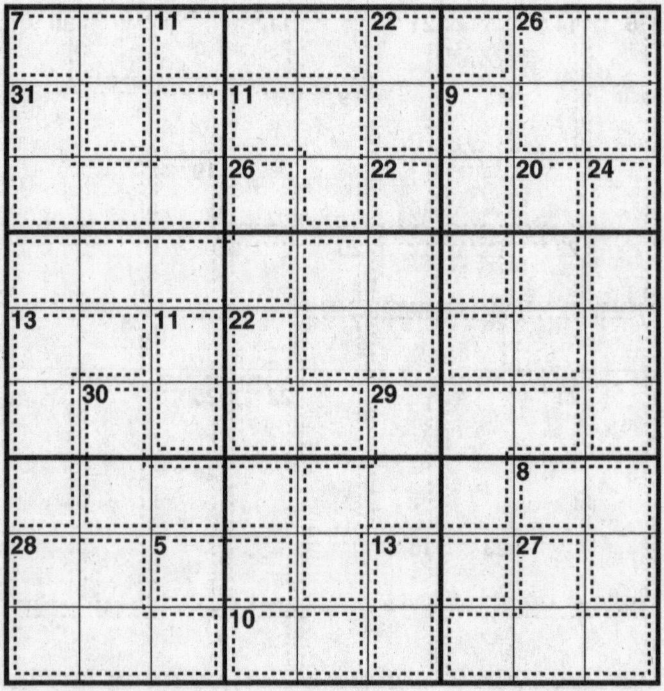

2 HOURS 15 MINUTES

TIME TAKEN............................

113

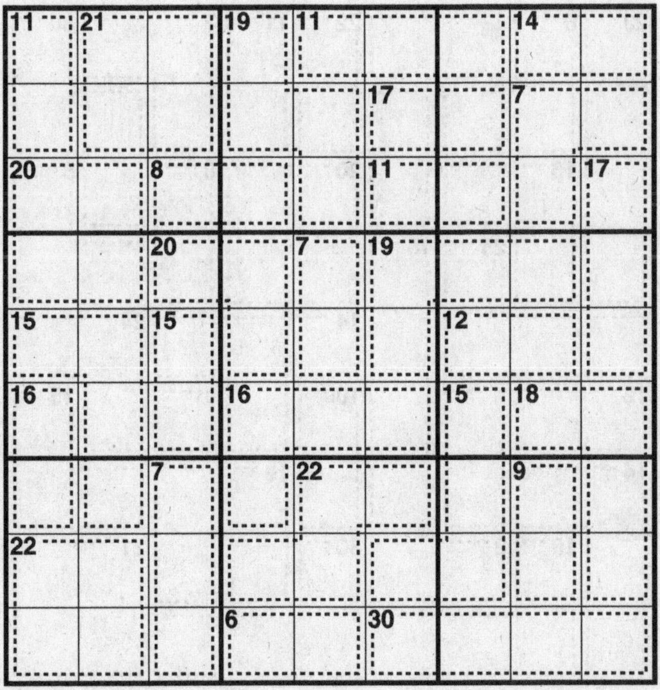

2 HOURS 15 MINUTES

TIME TAKEN..........................

114

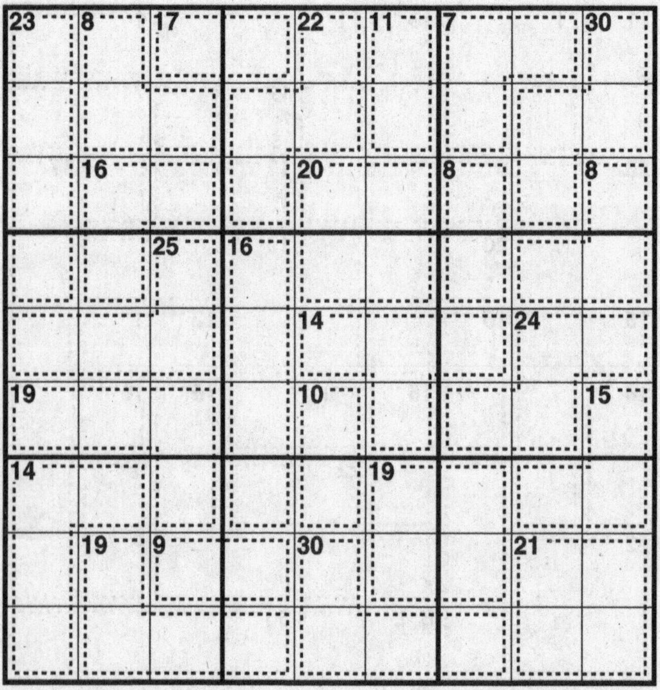

2 HOURS 25 MINUTES

TIME TAKEN..........................

115

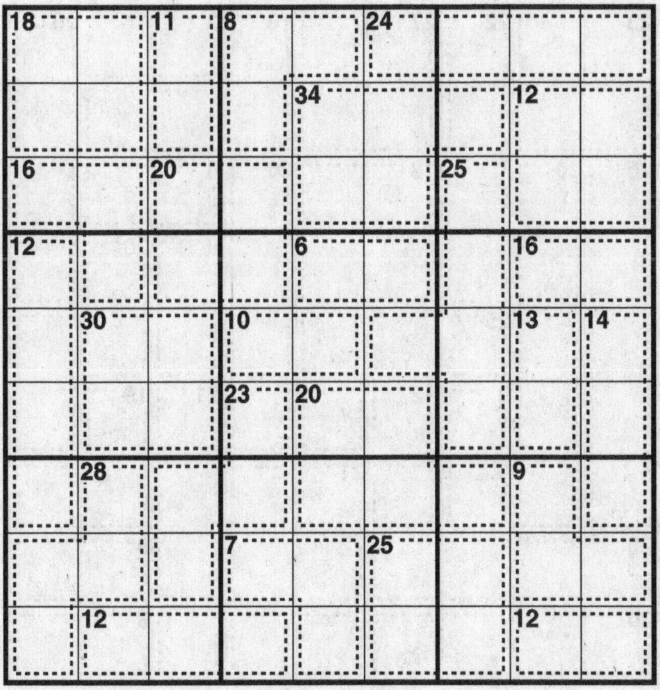

2 HOURS 25 MINUTES

TIME TAKEN.............................

116

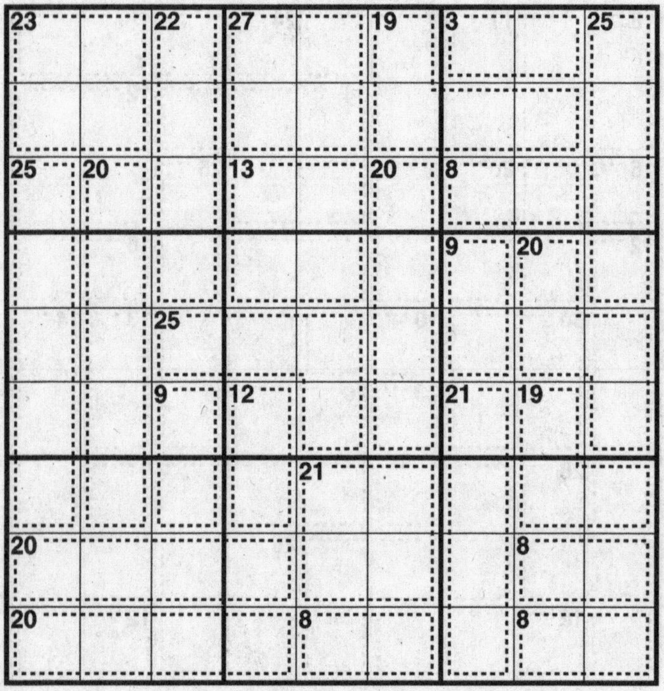

2 HOURS 30 MINUTES

TIME TAKEN........................

117

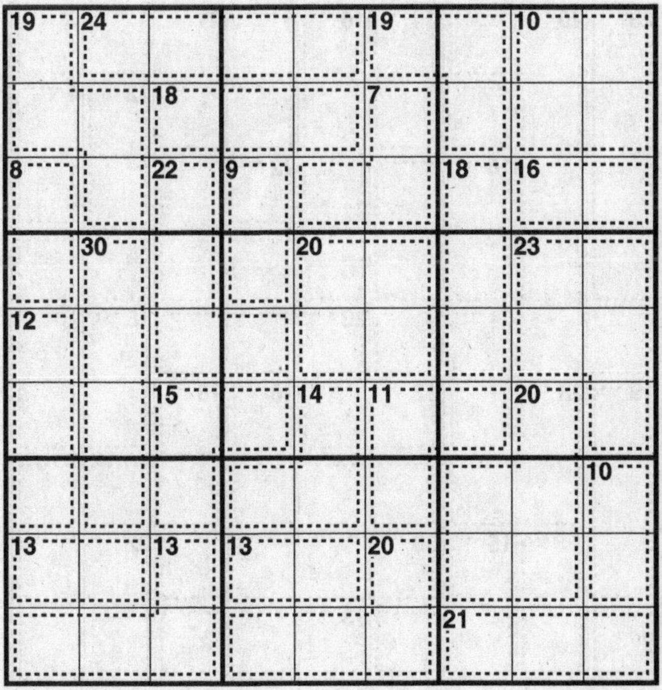

2 HOURS 30 MINUTES

TIME TAKEN............................

Extra Deadly

118

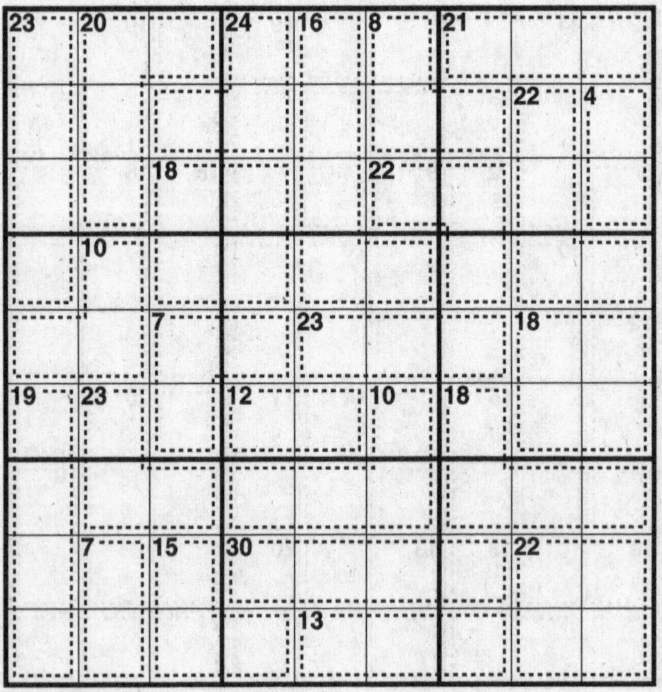

2 HOURS 30 MINUTES

TIME TAKEN.........................

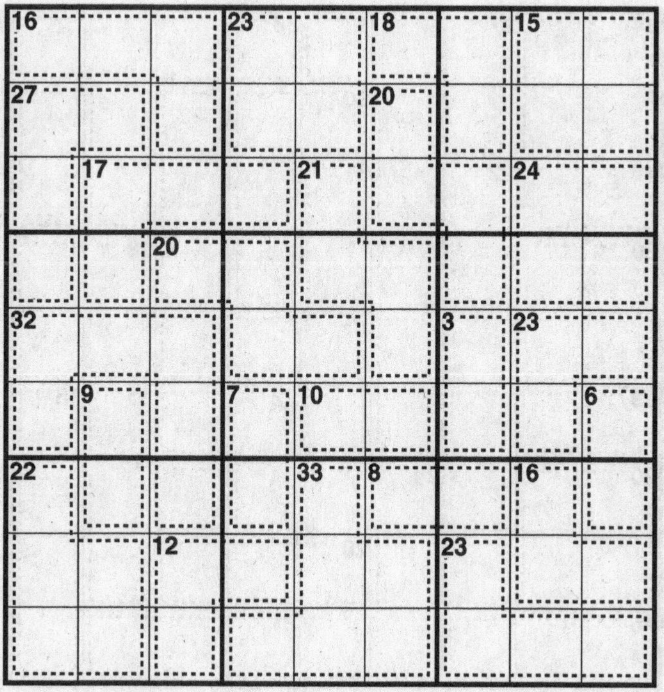

2 HOURS 30 MINUTES

TIME TAKEN..........................

Extra Deadly

120

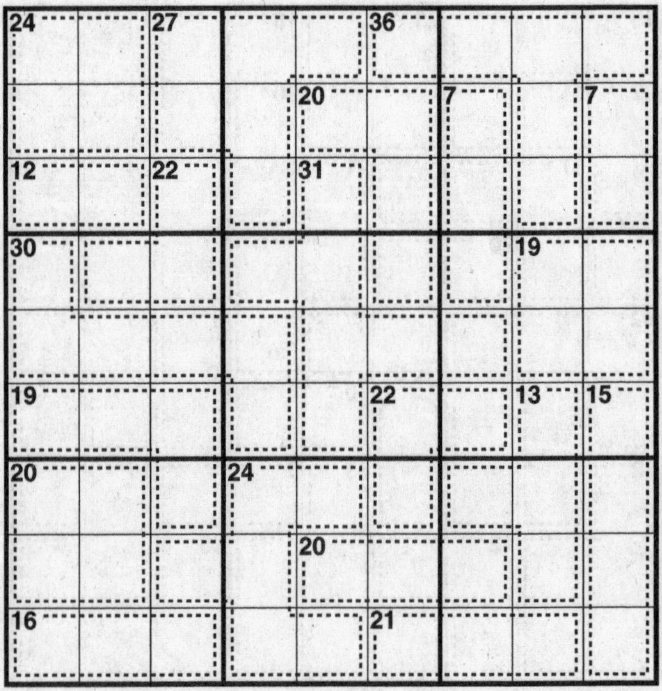

2 HOURS 30 MINUTES

TIME TAKEN............................

PUZZLES
Book Three

1

45 MINUTES

TIME TAKEN.............................

Deadly

2

45 MINUTES

TIME TAKEN.................................

3

45 MINUTES

TIME TAKEN..............................

Deadly

4

45 MINUTES

TIME TAKEN..............................

5

45 MINUTES

TIME TAKEN..............................

Deadly

6

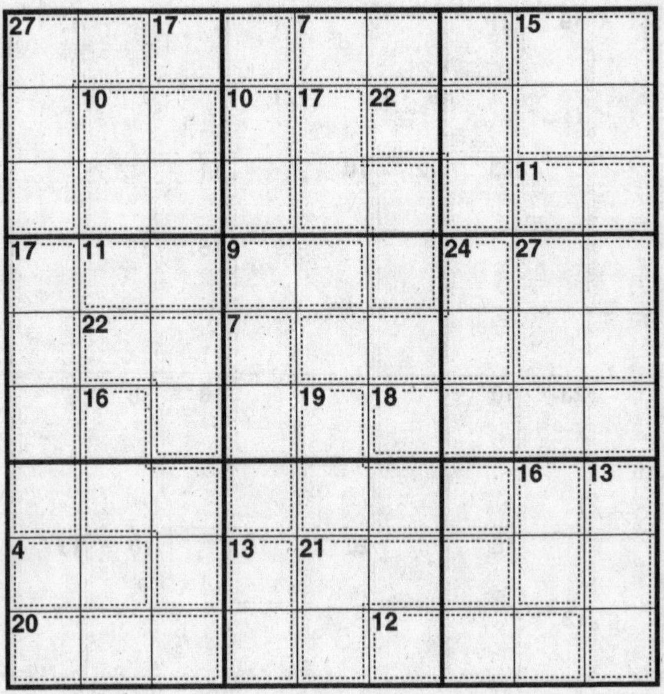

50 MINUTES

TIME TAKEN..............................

7

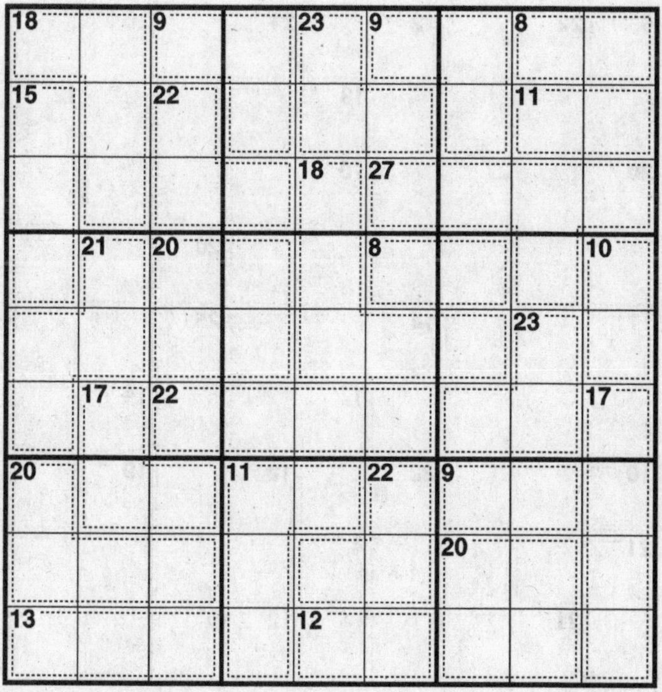

50 MINUTES

TIME TAKEN..............................

Deadly

8

50 MINUTES

TIME TAKEN............................

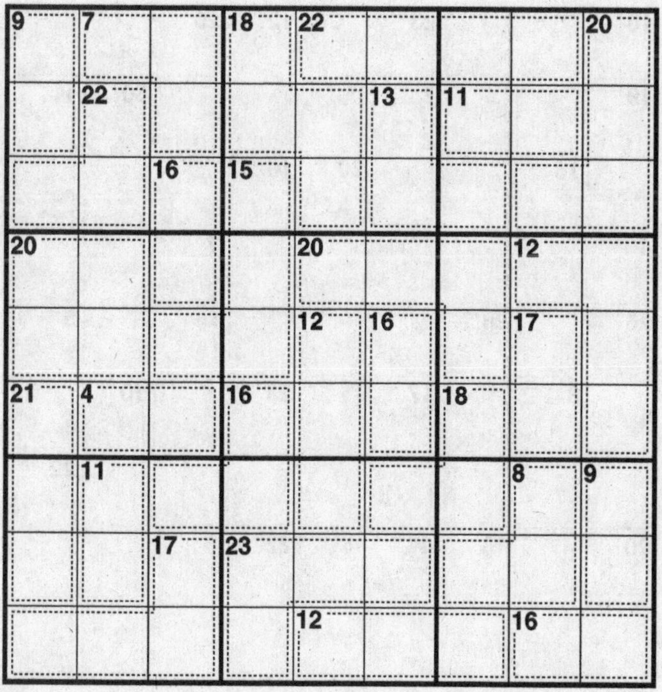

50 MINUTES

TIME TAKEN..............................

Deadly

10

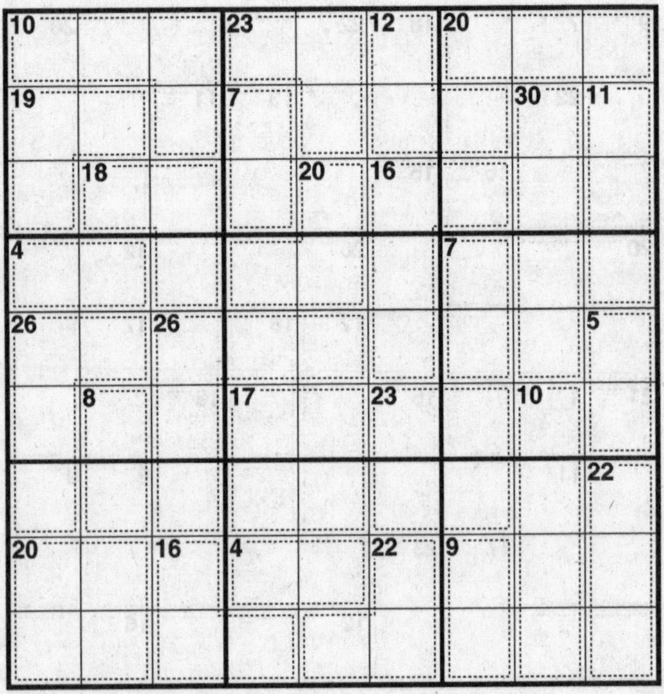

50 MINUTES

TIME TAKEN.................................

50 MINUTES

TIME TAKEN..............................

Deadly

12

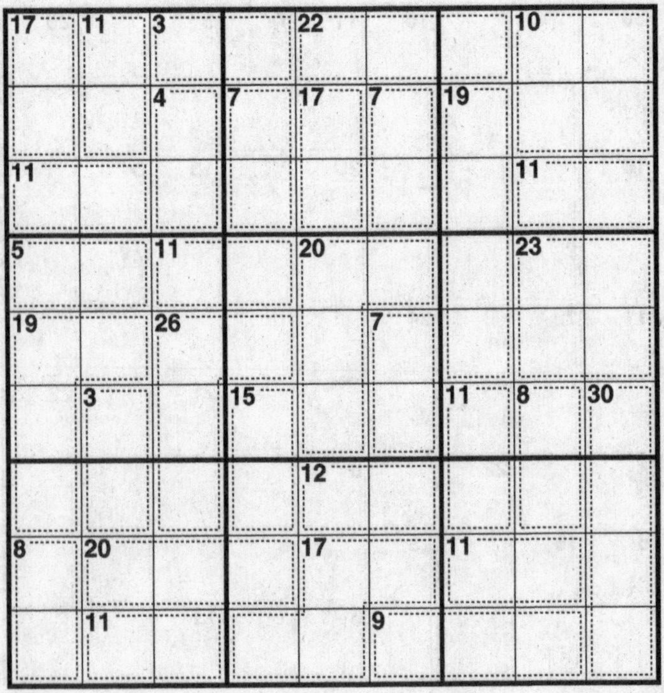

50 MINUTES

TIME TAKEN..............................

13

50 MINUTES

TIME TAKEN............................

Deadly

14

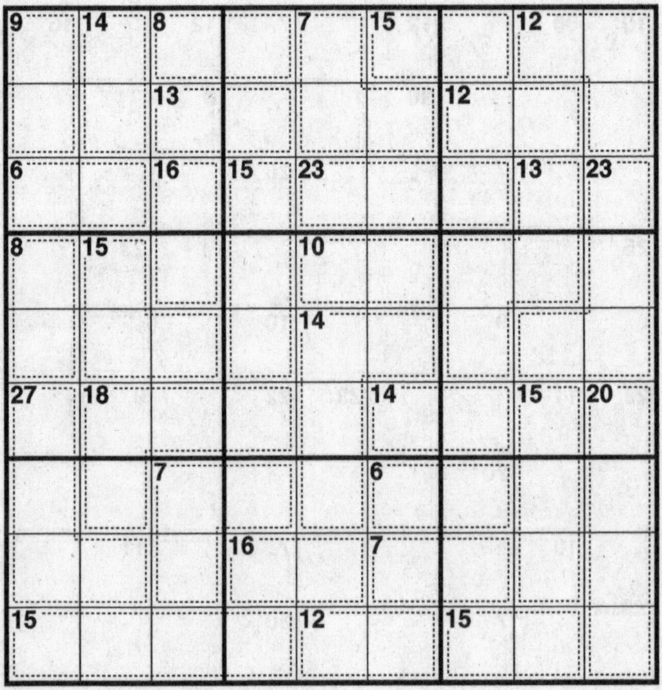

50 MINUTES

TIME TAKEN..............................

15

50 MINUTES

TIME TAKEN............................

Deadly

16

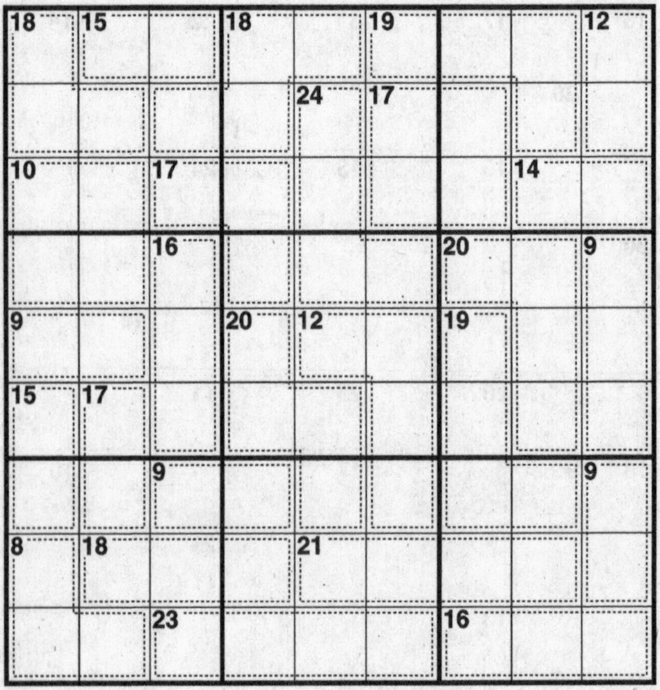

50 MINUTES

TIME TAKEN..............................

17

50 MINUTES

TIME TAKEN..............................

Deadly

18

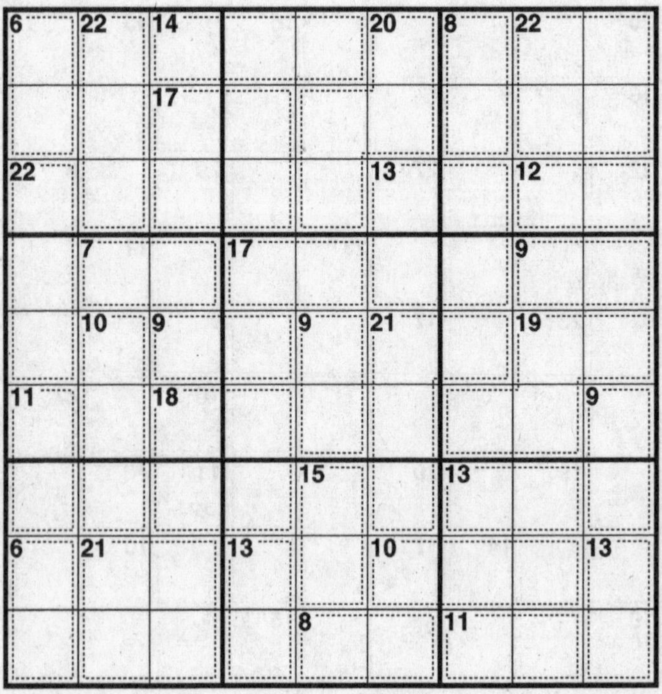

50 MINUTES

TIME TAKEN..............................

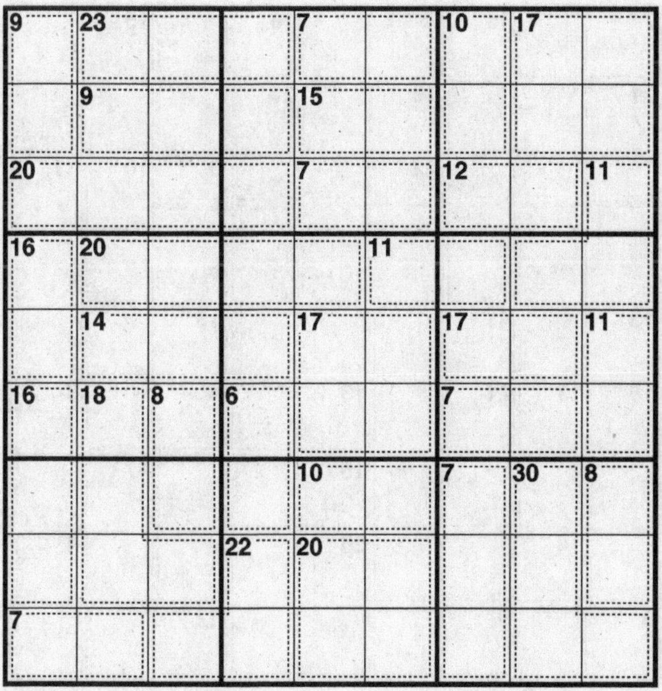

50 MINUTES

TIME TAKEN..............................

20

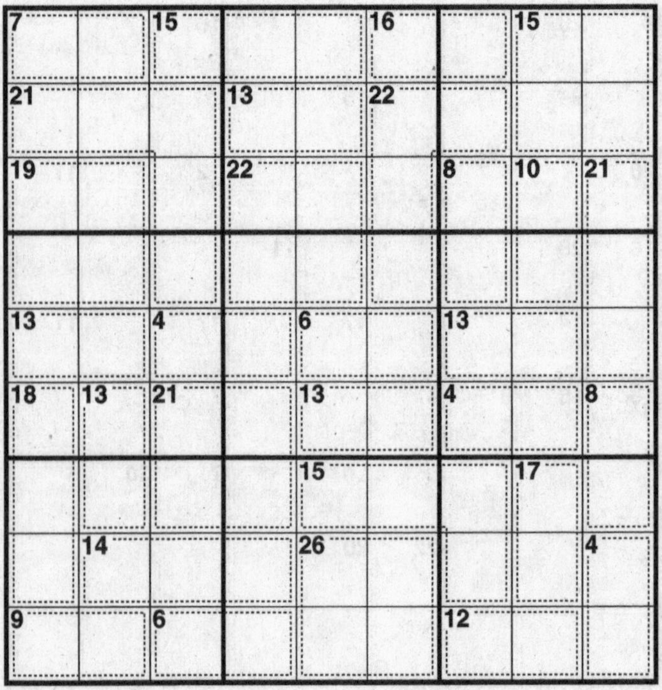

50 MINUTES

TIME TAKEN.............................

21

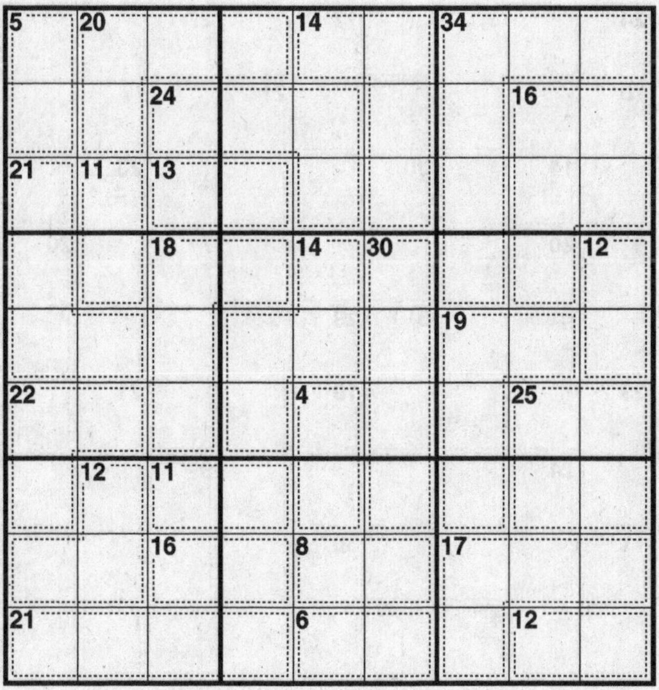

52 MINUTES

TIME TAKEN..............................

Deadly

22

52 MINUTES

TIME TAKEN..............................

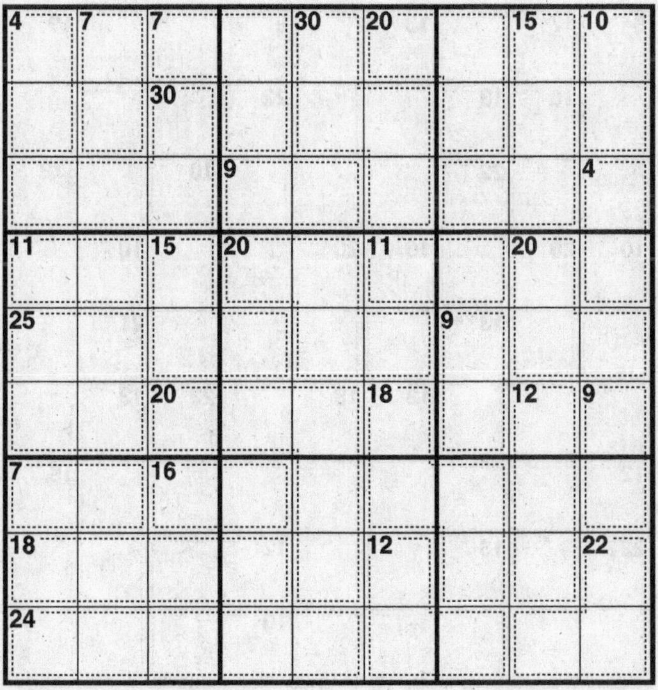

52 MINUTES

TIME TAKEN..............................

24

52 MINUTES

TIME TAKEN..............................

25

52 MINUTES

TIME TAKEN..............................

Deadly

26

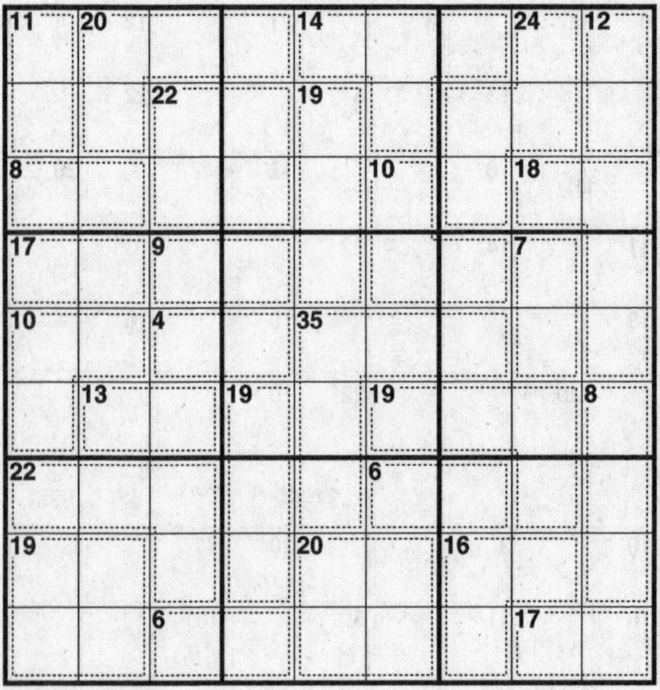

55 MINUTES

TIME TAKEN............................

55 MINUTES

TIME TAKEN..................................

Deadly

28

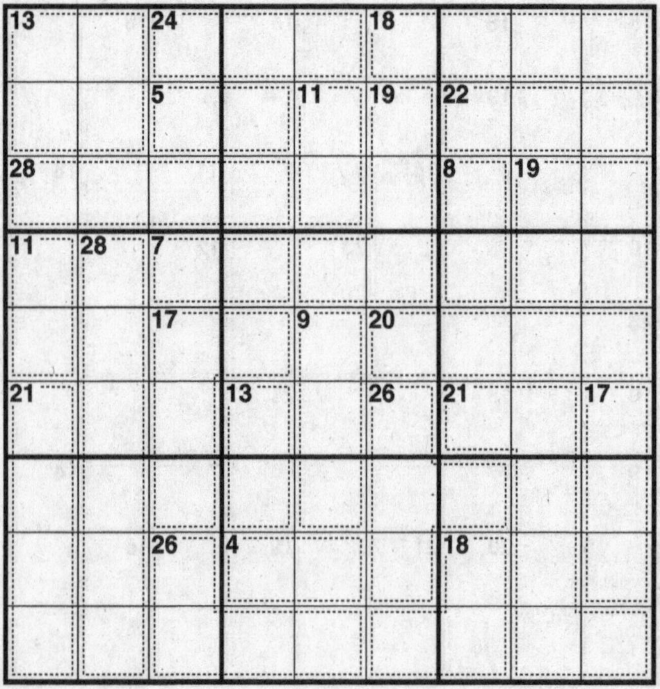

55 MINUTES

TIME TAKEN..............................

55 MINUTES

TIME TAKEN..................................

Deadly

30

55 MINUTES

TIME TAKEN..............................

55 MINUTES

TIME TAKEN.............................

32

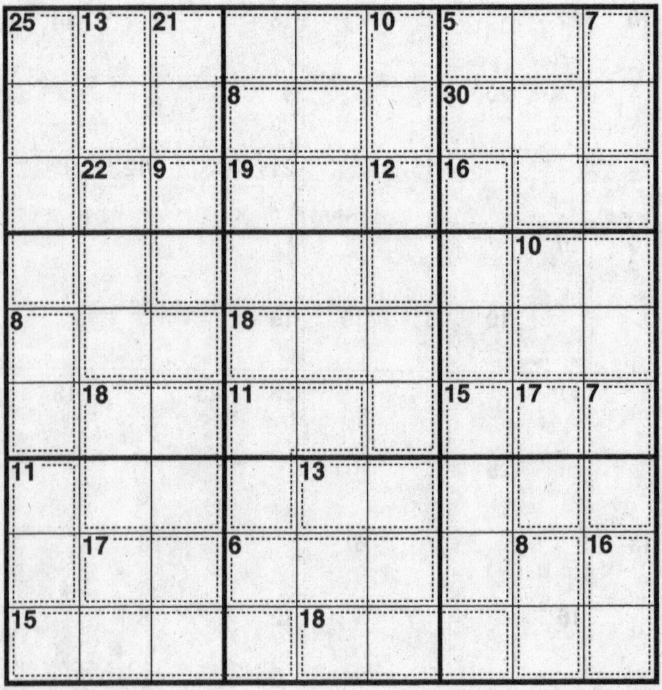

55 MINUTES

TIME TAKEN..............................

33

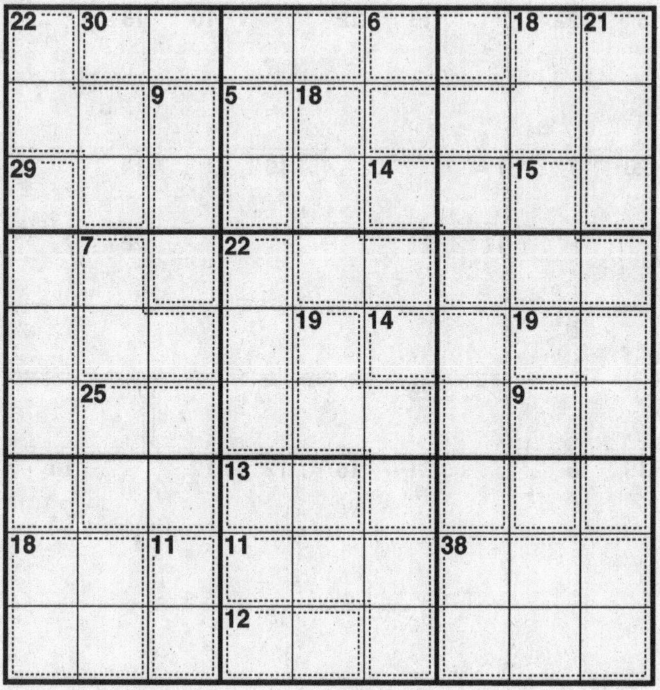

58 MINUTES

TIME TAKEN..............................

Deadly

34

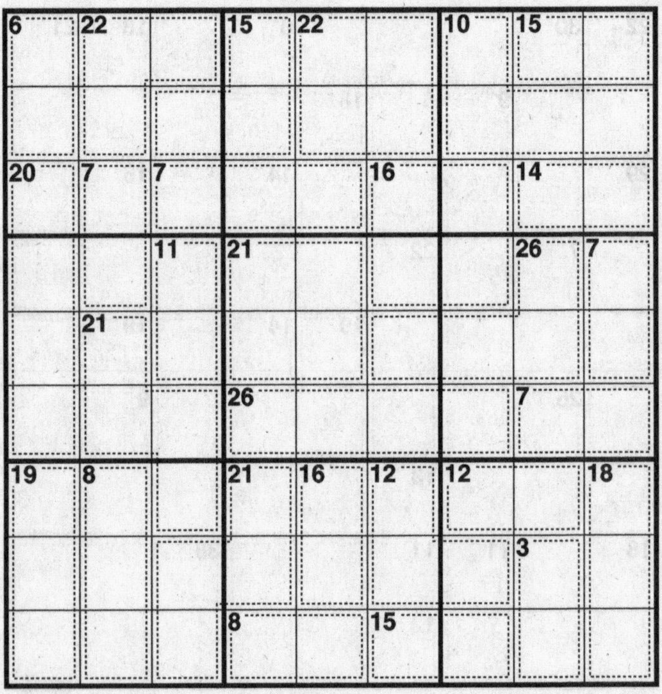

58 MINUTES

TIME TAKEN..............................

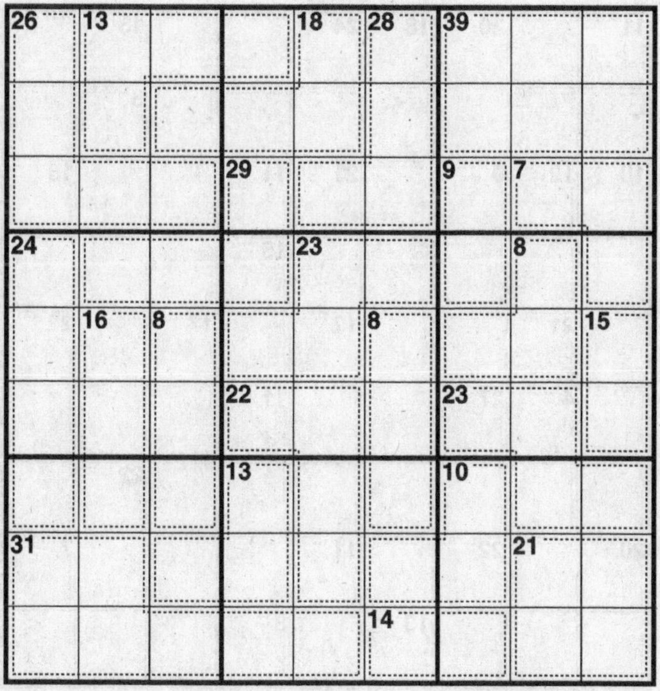

58 MINUTES

TIME TAKEN..............................

36

58 MINUTES

TIME TAKEN..............................

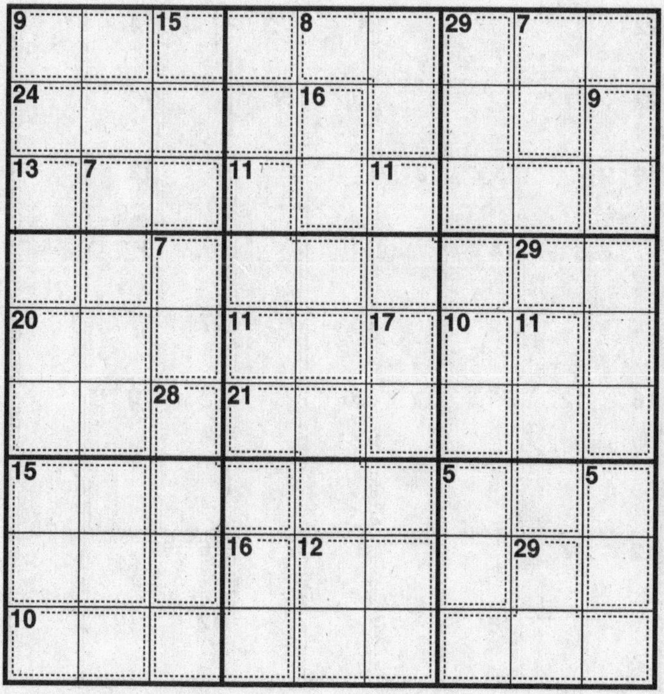

58 MINUTES

TIME TAKEN..............................

Deadly

38

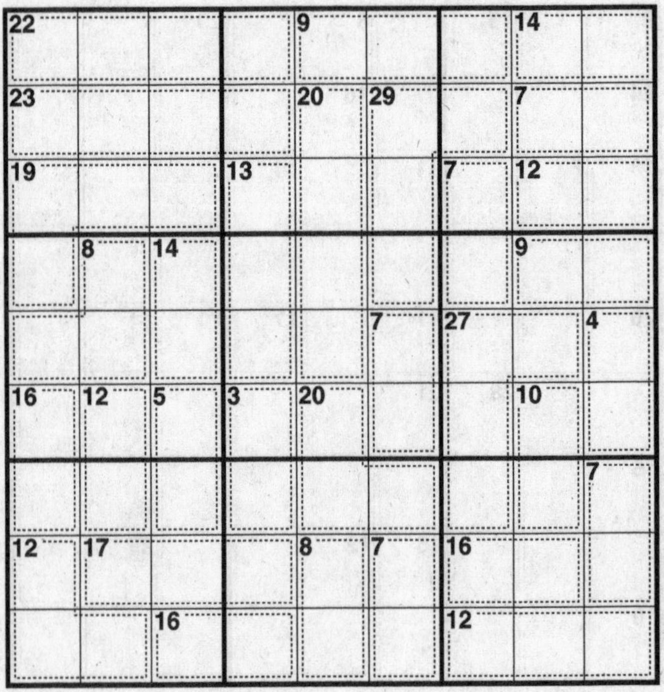

58 MINUTES

TIME TAKEN..............................

39

1 HOUR

TIME TAKEN..............................

Deadly

40

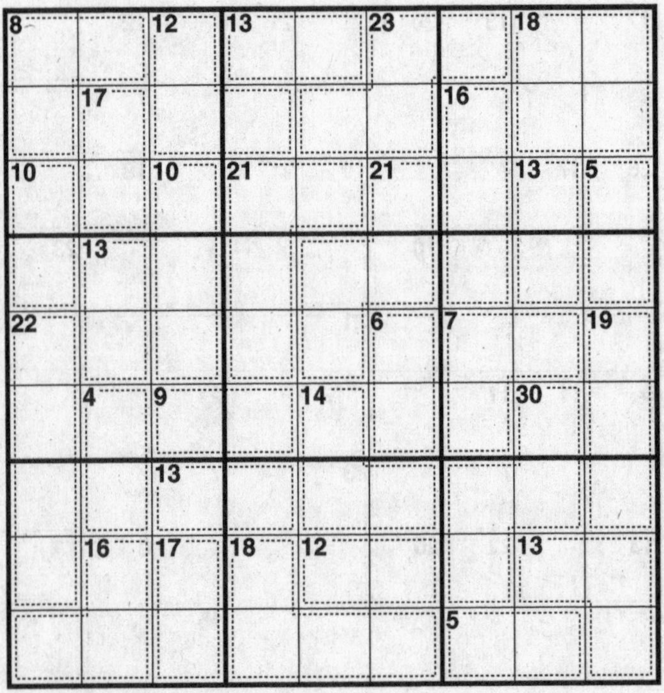

1 HOUR

TIME TAKEN.............................

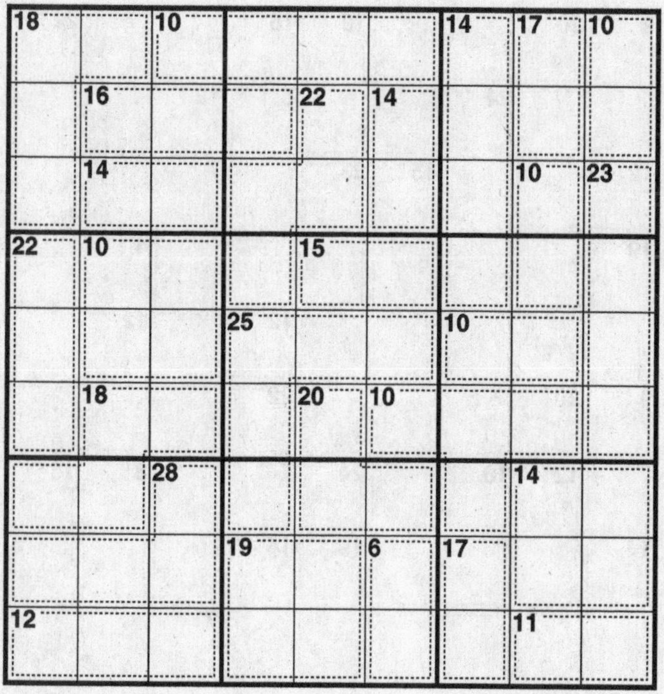

1 HOUR

TIME TAKEN..................................

42

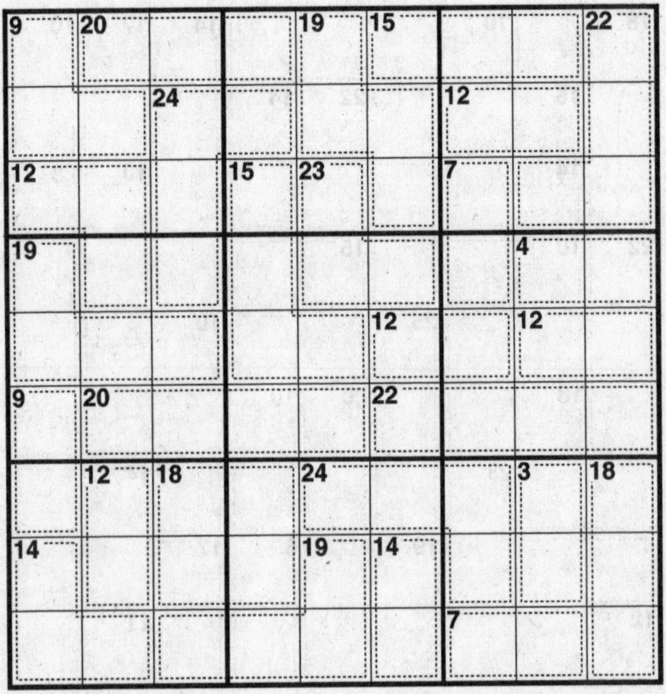

1 HOUR

TIME TAKEN..............................

43

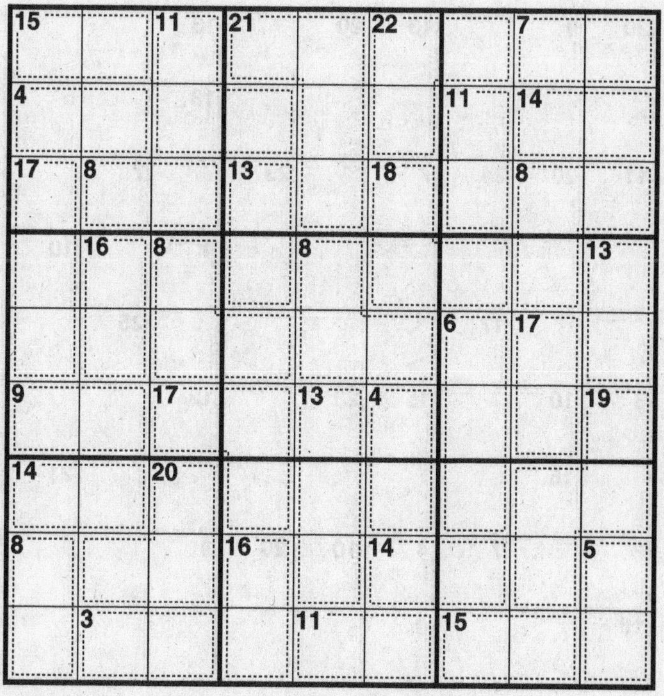

1 HOUR

TIME TAKEN..............................

Deadly

44

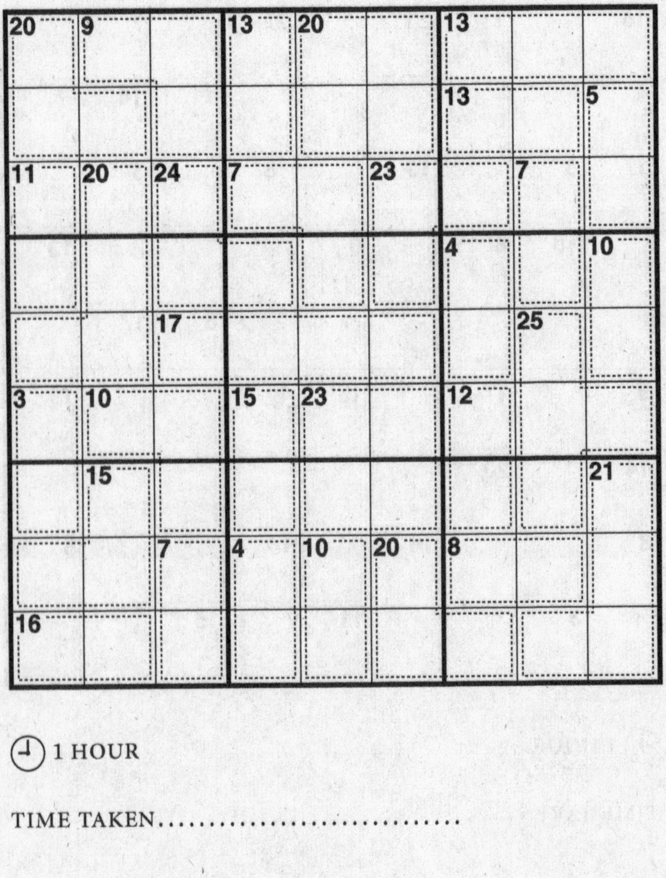

1 HOUR

TIME TAKEN.............................

45

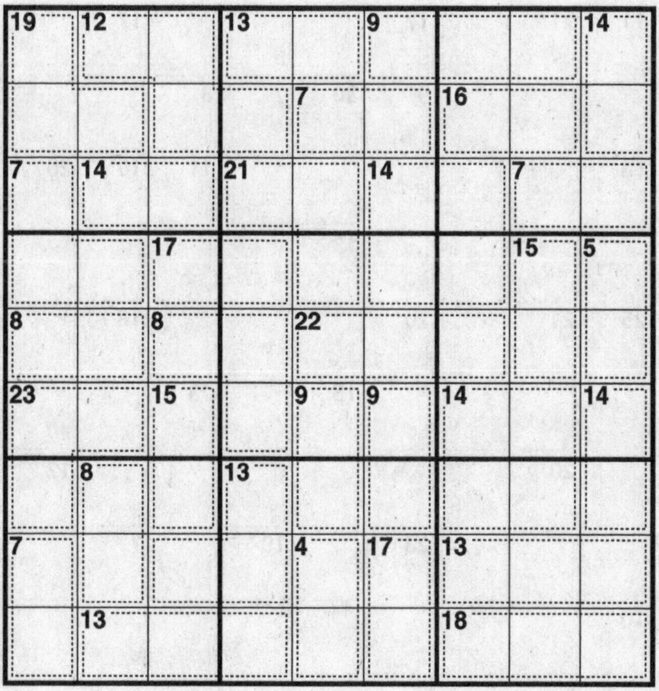

1 HOUR

TIME TAKEN..............................

1 HOUR

TIME TAKEN..............................

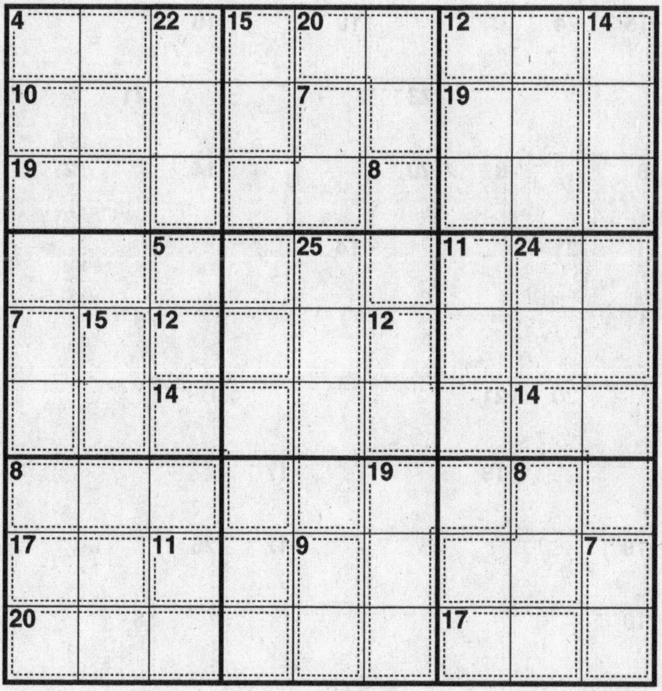

1 HOUR

TIME TAKEN...............................

48

1 HOUR

TIME TAKEN............................

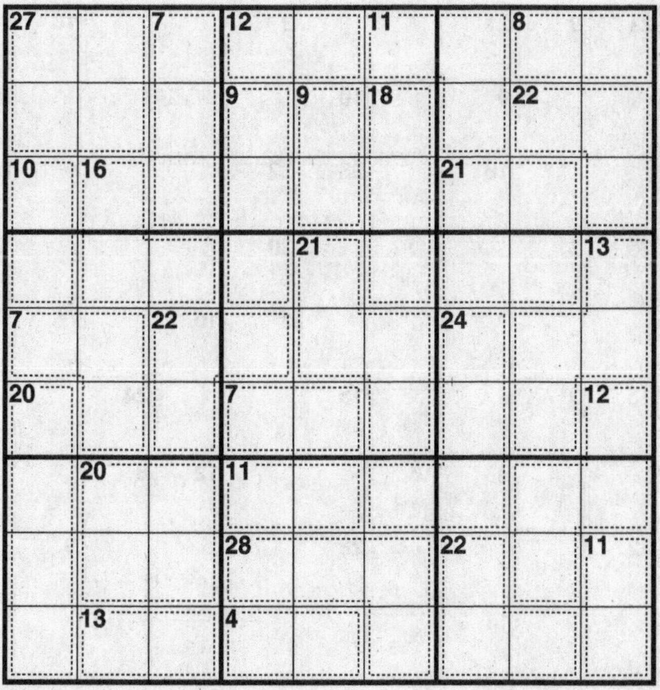

1 HOUR

TIME TAKEN..............................

50

1 HOUR

TIME TAKEN..............................

51

1 HOUR

TIME TAKEN.............................

Deadly

52

1 HOUR

TIME TAKEN..............................

53

1 HOUR

TIME TAKEN..............................

Deadly

54

1 HOUR

TIME TAKEN..............................

55

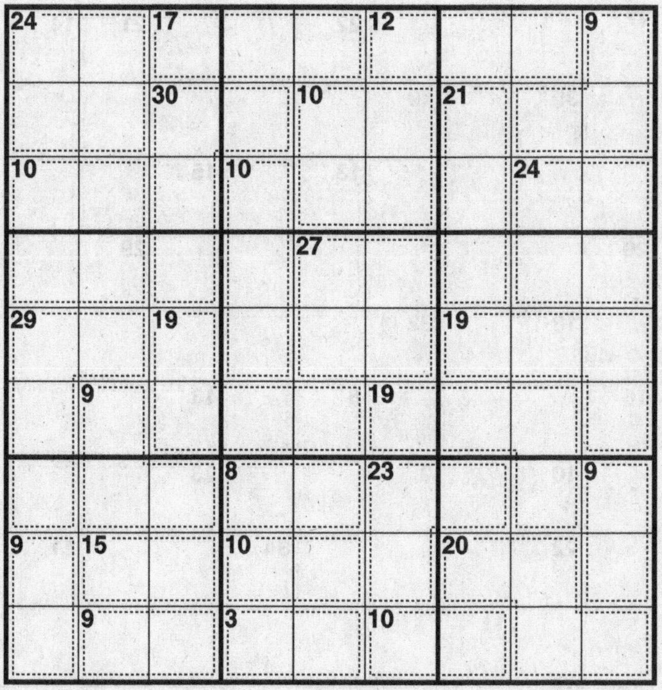

1 HOUR

TIME TAKEN..............................

Deadly

56

1 HOUR

TIME TAKEN..............................

57

1 HOUR

TIME TAKEN..............................

Deadly

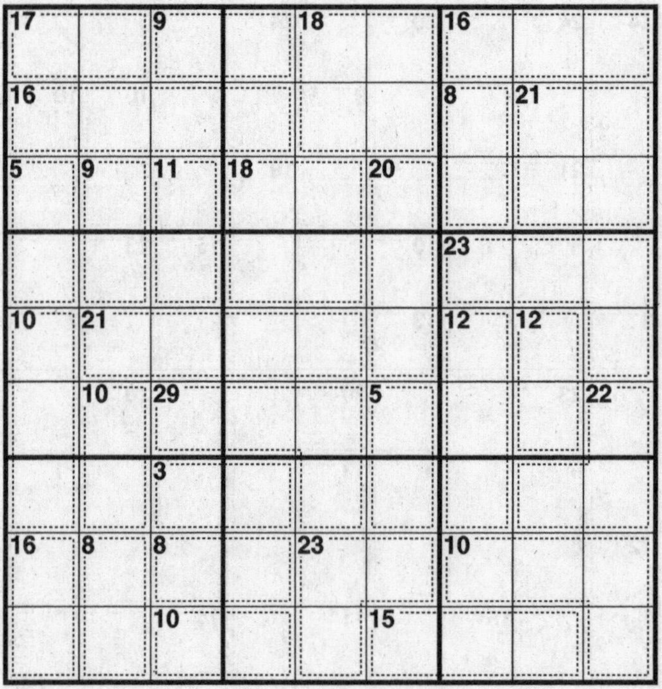

1 HOUR

TIME TAKEN..............................

59

1 HOUR

TIME TAKEN..............................

Deadly

60

1 HOUR 5 MINUTES

TIME TAKEN..............................

61

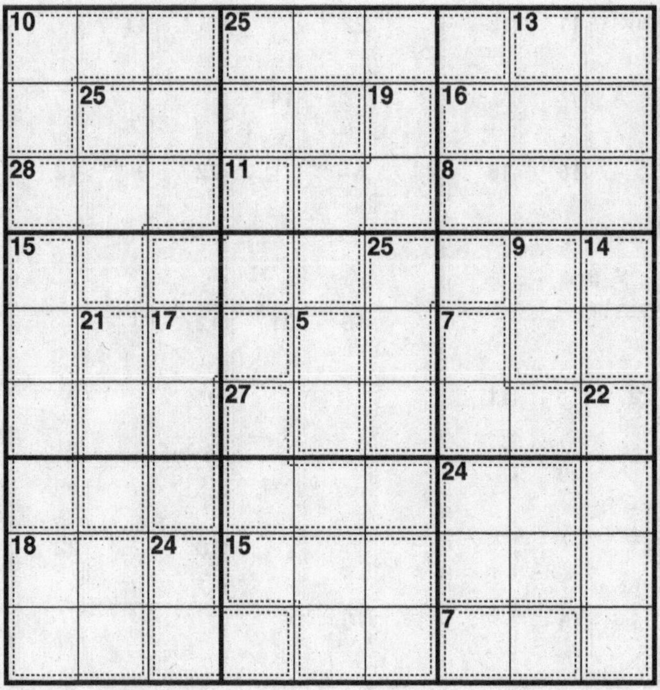

1 HOUR 5 MINUTES

TIME TAKEN..............................

1 HOUR 5 MINUTES

TIME TAKEN..............................

1 HOUR 5 MINUTES

TIME TAKEN..............................

64

1 HOUR 5 MINUTES

TIME TAKEN..............................

65

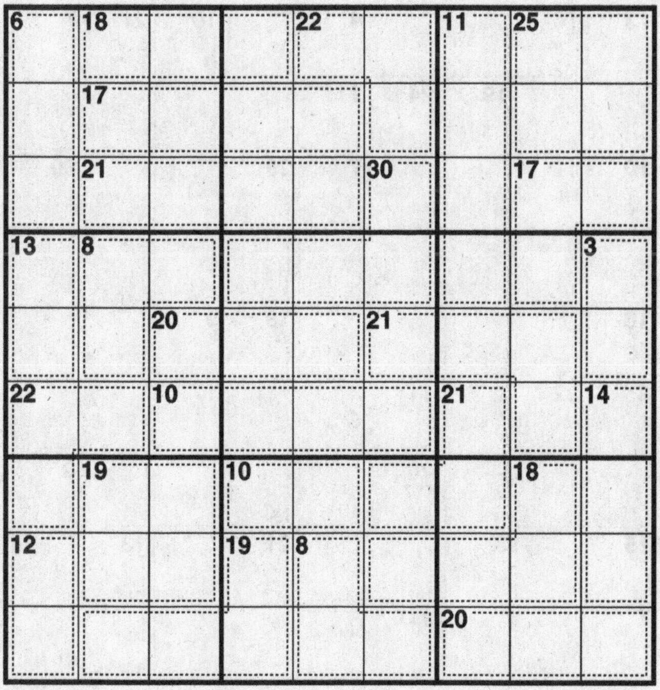

1 HOUR 5 MINUTES

TIME TAKEN..............................

Deadly

66

1 HOUR 5 MINUTES

TIME TAKEN..............................

67

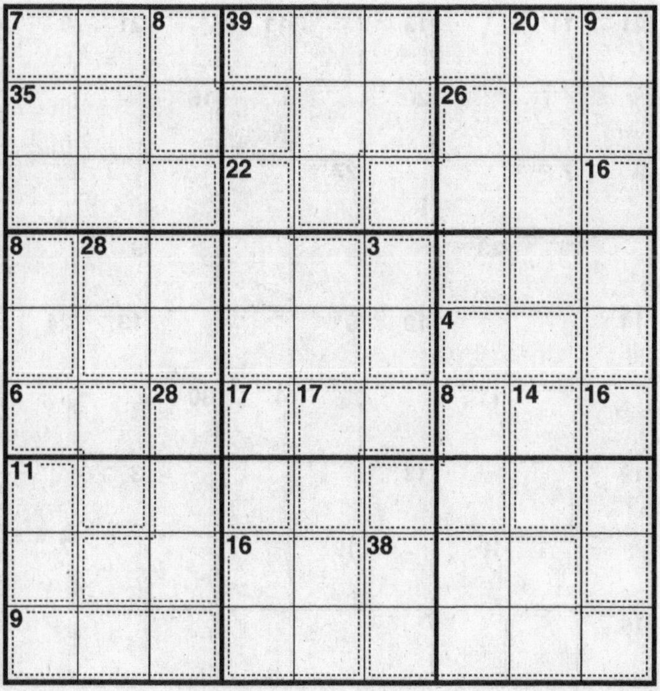

1 HOUR 5 MINUTES

TIME TAKEN..............................

Deadly

68

1 HOUR 5 MINUTES

TIME TAKEN.................................

1 HOUR 5 MINUTES

TIME TAKEN..................................

70

1 HOUR 5 MINUTES

TIME TAKEN..............................

1 HOUR 10 MINUTES

TIME TAKEN..............................

72

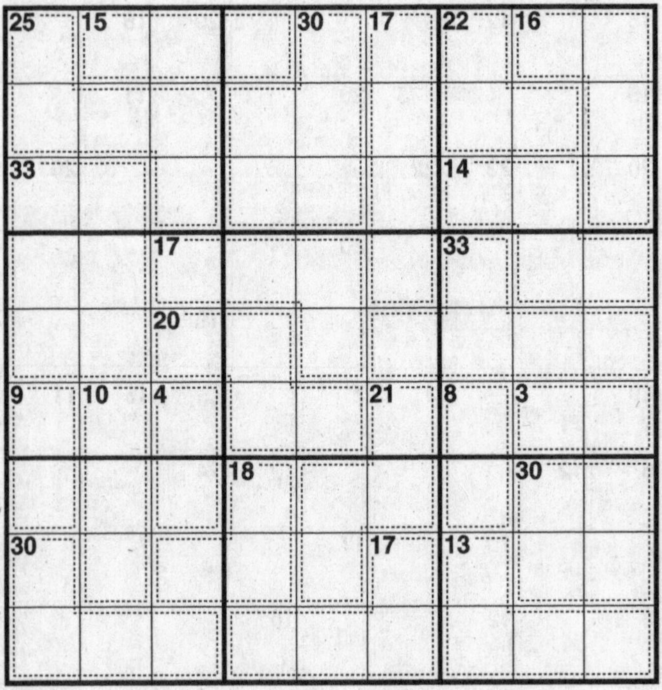

1 HOUR 10 MINUTES

TIME TAKEN.............................

1 HOUR 10 MINUTES

TIME TAKEN..............................

74

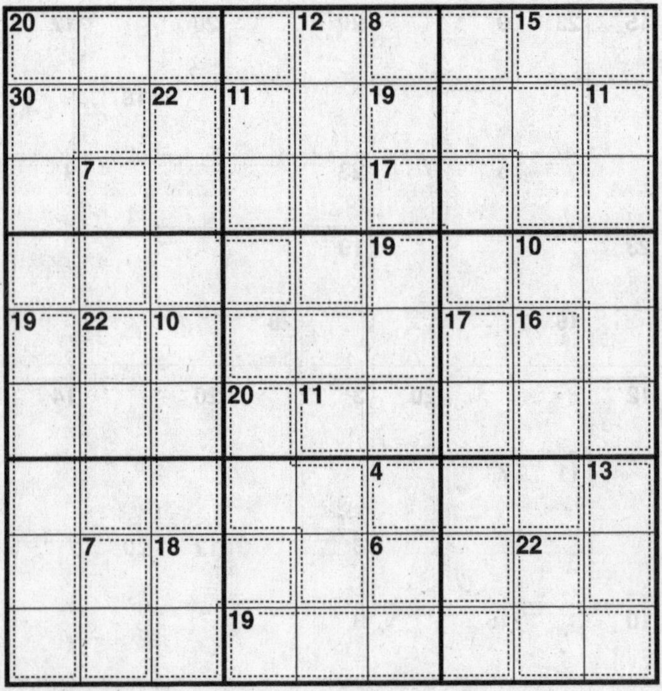

1 HOUR 10 MINUTES

TIME TAKEN..............................

75

1 HOUR 10 MINUTES

TIME TAKEN..............................

Deadly

76

1 HOUR 10 MINUTES

TIME TAKEN............................

77

1 HOUR 10 MINUTES

TIME TAKEN.............................

Deadly

78

1 HOUR 10 MINUTES

TIME TAKEN..............................

79

1 HOUR 10 MINUTES

TIME TAKEN..............................

Deadly

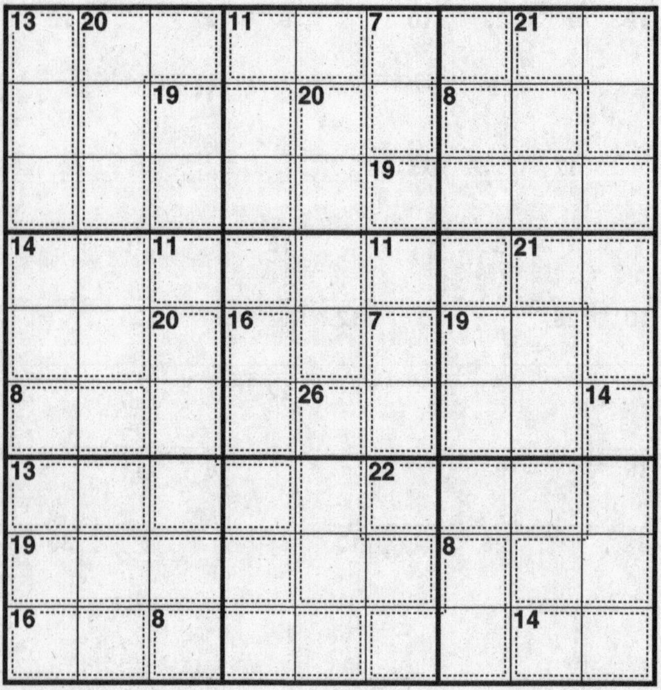

1 HOUR 10 MINUTES

TIME TAKEN..............................

81

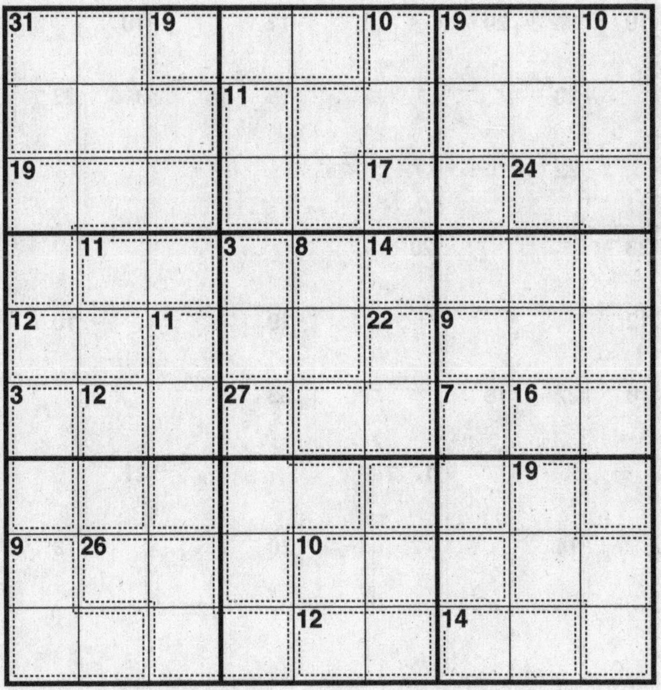

1 HOUR 15 MINUTES

TIME TAKEN..............................

Deadly

82

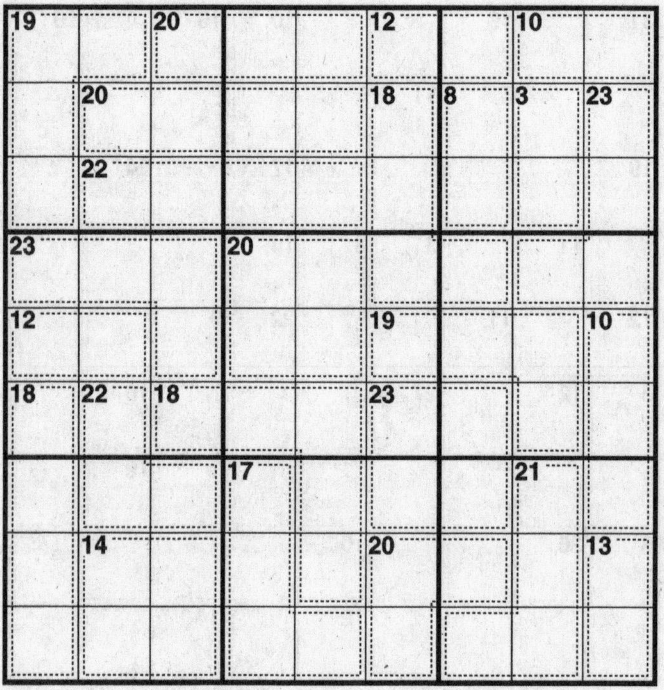

1 HOUR 15 MINUTES

TIME TAKEN..............................

83

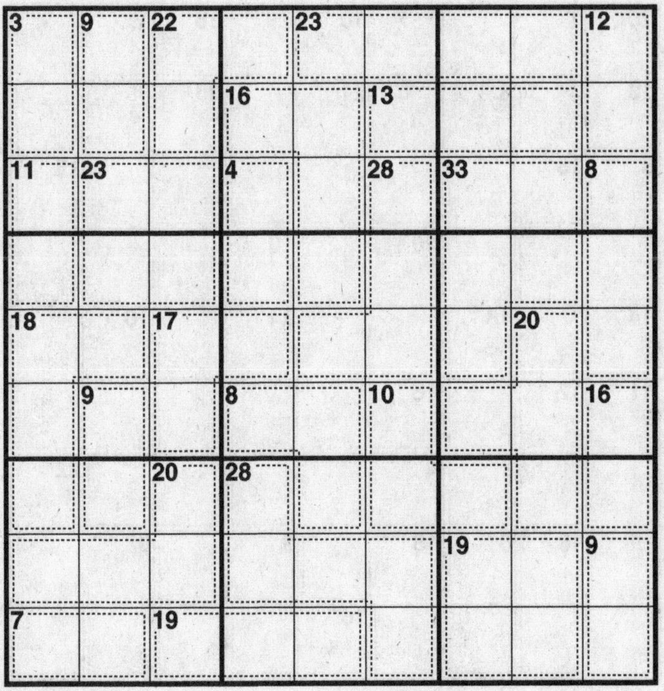

1 HOUR 15 MINUTES

TIME TAKEN..............................

Deadly

84

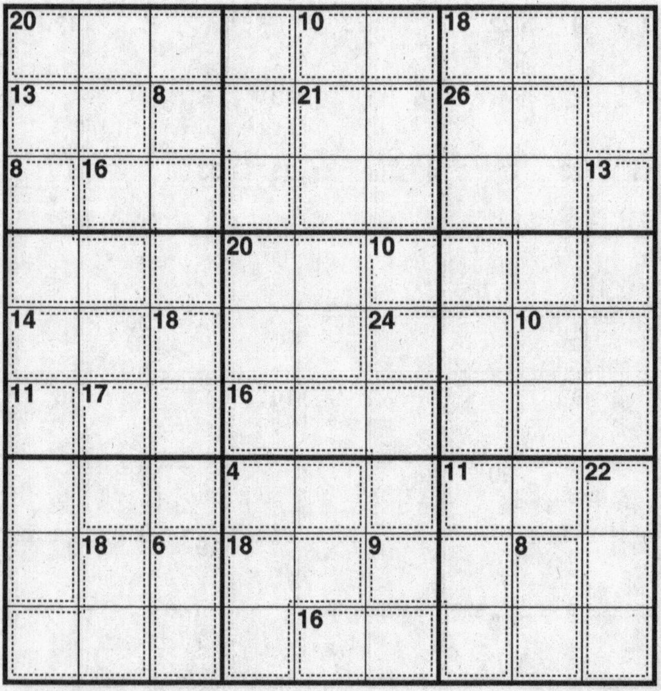

1 HOUR 15 MINUTES

TIME TAKEN..............................

1 HOUR 15 MINUTES

TIME TAKEN..............................

Deadly

86

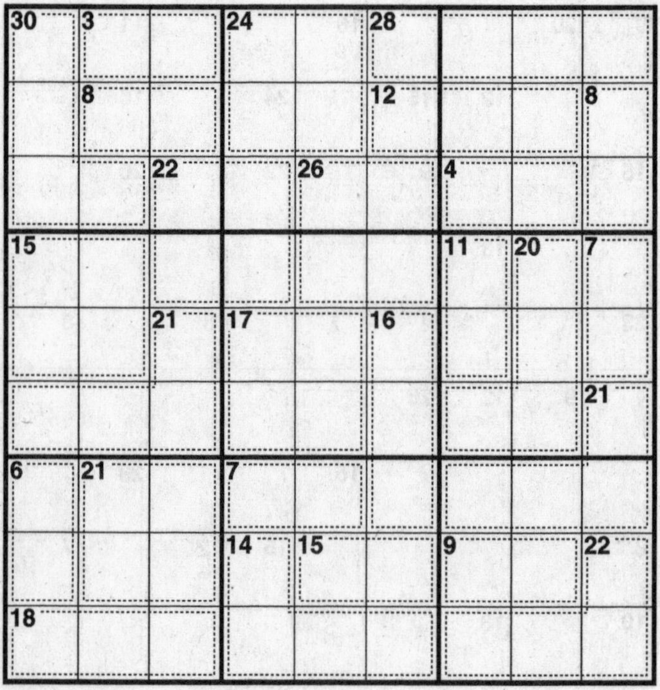

1 HOUR 15 MINUTES

TIME TAKEN..............................

1 HOUR 20 MINUTES

TIME TAKEN..............................

88

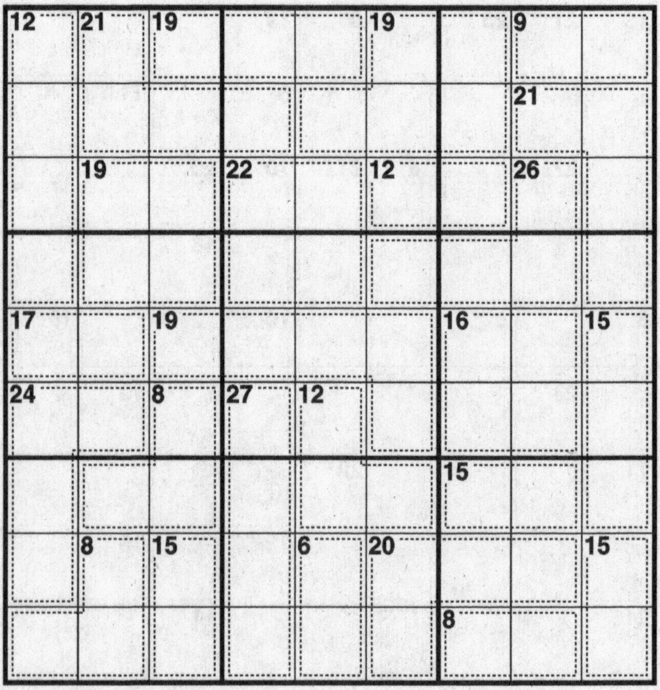

1 HOUR 20 MINUTES

TIME TAKEN..............................

89

1 HOUR 20 MINUTES

TIME TAKEN...............................

Deadly

90

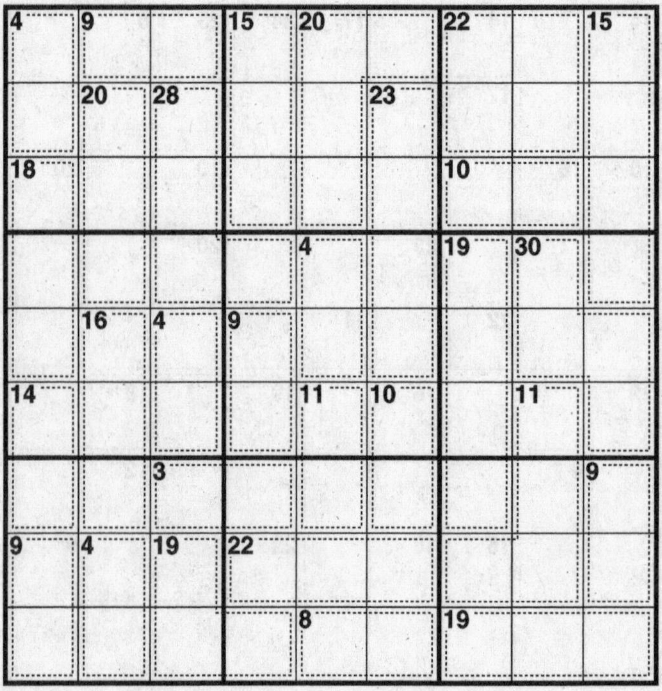

1 HOUR 20 MINUTES

TIME TAKEN..............................

91

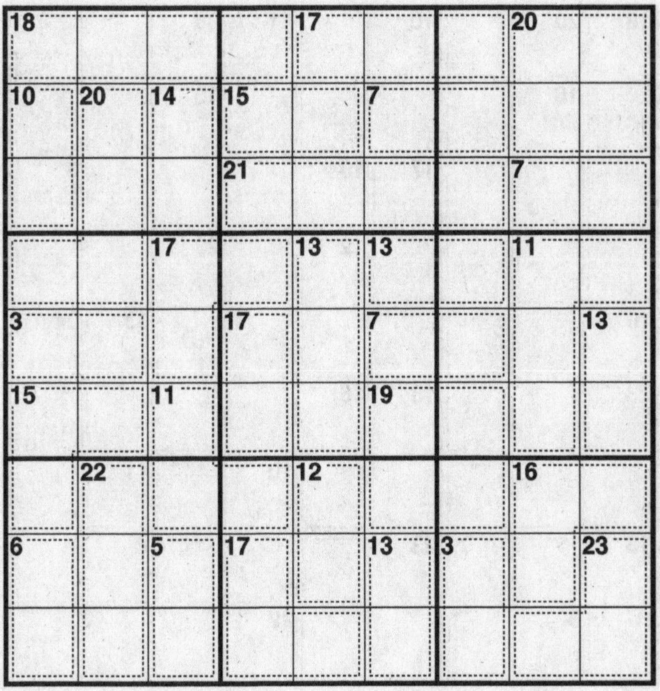

1 HOUR 20 MINUTES

TIME TAKEN..............................

Deadly

92

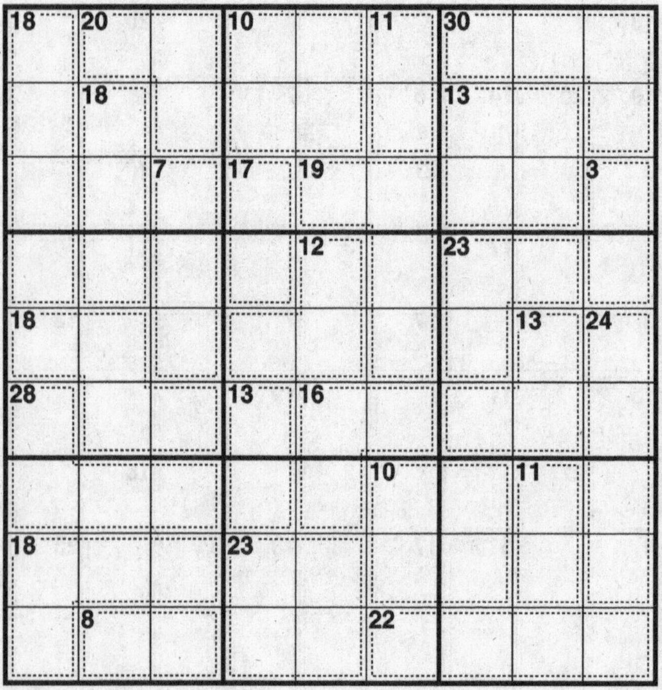

1 HOUR 20 MINUTES

TIME TAKEN..............................

1 HOUR 25 MINUTES

TIME TAKEN..............................

Deadly

1 HOUR 25 MINUTES

TIME TAKEN...............................

1 HOUR 25 MINUTES

TIME TAKEN..............................

96

1 HOUR 25 MINUTES

TIME TAKEN............................

1 HOUR 30 MINUTES

TIME TAKEN..............................

98

1 HOUR 30 MINUTES

TIME TAKEN..............................

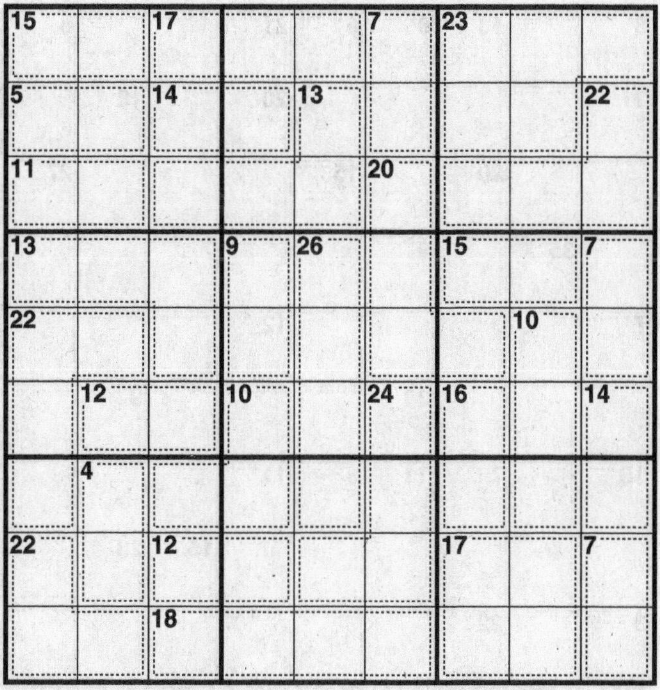

1 HOUR 30 MINUTES

TIME TAKEN..............................

Deadly

100

1 HOUR 30 MINUTES

TIME TAKEN..............................

101

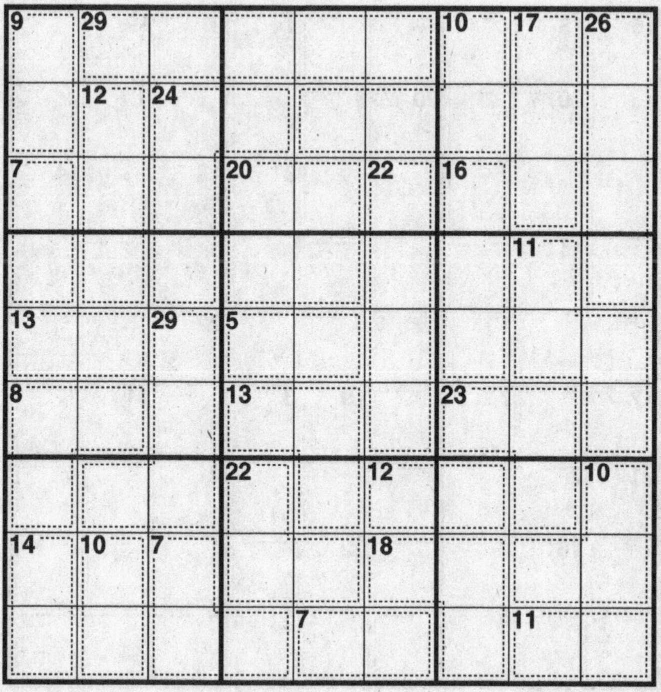

1 HOUR 45 MINUTES

TIME TAKEN.............................

Extra Deadly

102

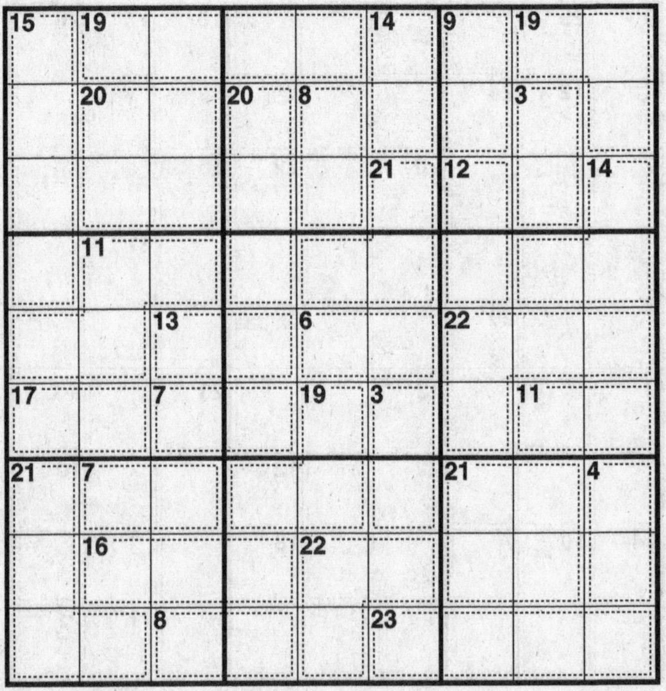

1 HOUR 45 MINUTES

TIME TAKEN............................

1 HOUR 45 MINUTES

TIME TAKEN.............................

Extra Deadly

104

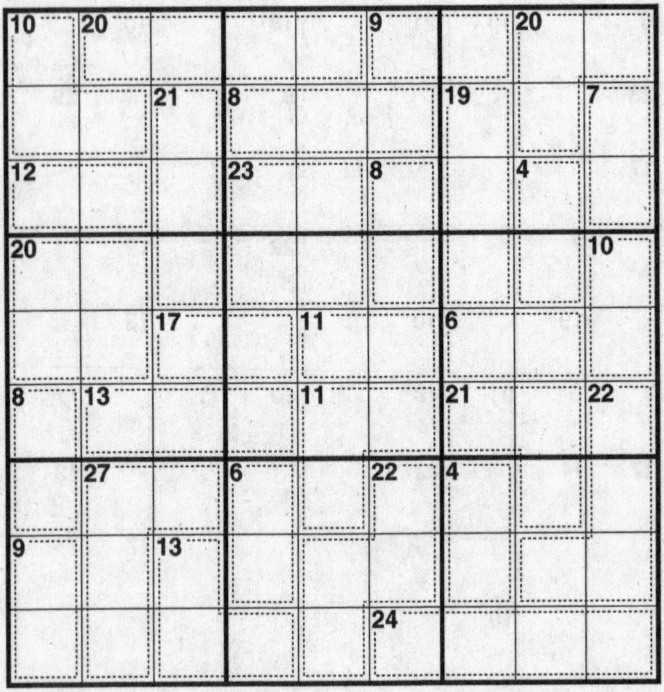

1 HOUR 45 MINUTES

TIME TAKEN...............................

105

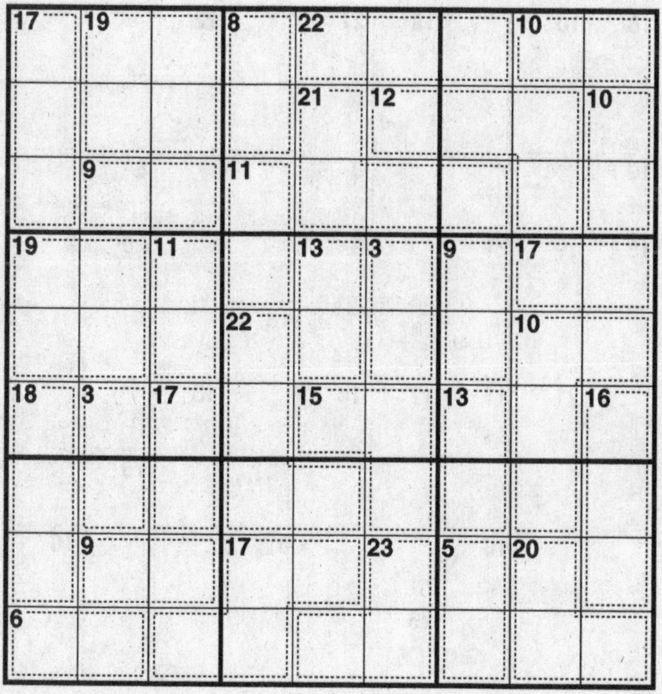

1 HOUR 45 MINUTES

TIME TAKEN..............................

Extra Deadly

106

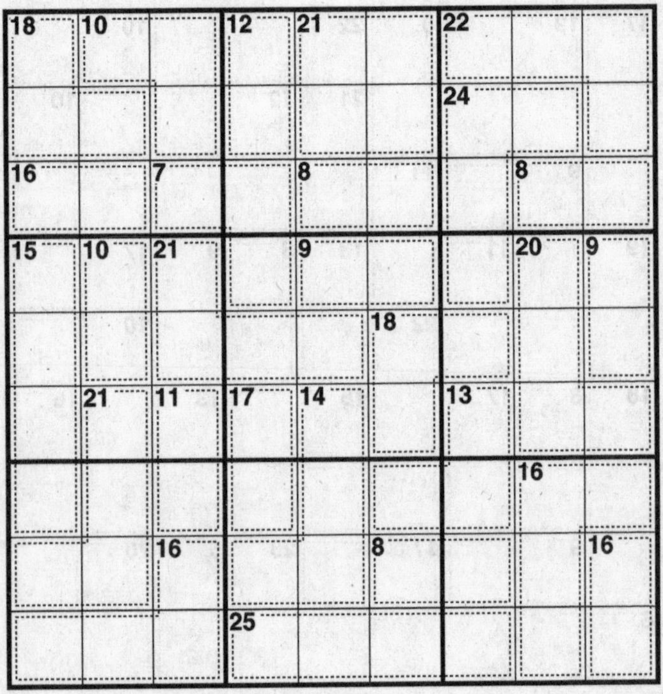

2 HOURS

TIME TAKEN.............................

107

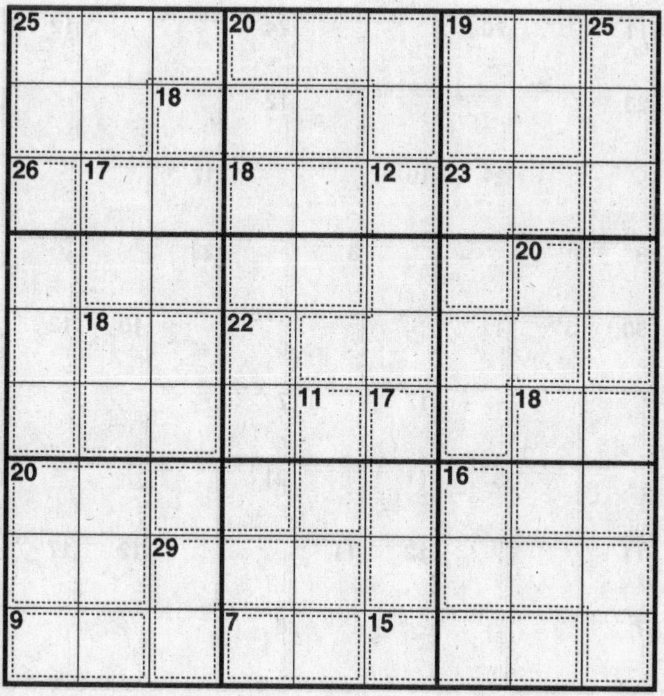

2 HOURS

TIME TAKEN..............................

Extra Deadly

108

2 HOURS

TIME TAKEN............................

2 HOURS

TIME TAKEN..............................

Extra Deadly

110

2 HOURS

TIME TAKEN..............................

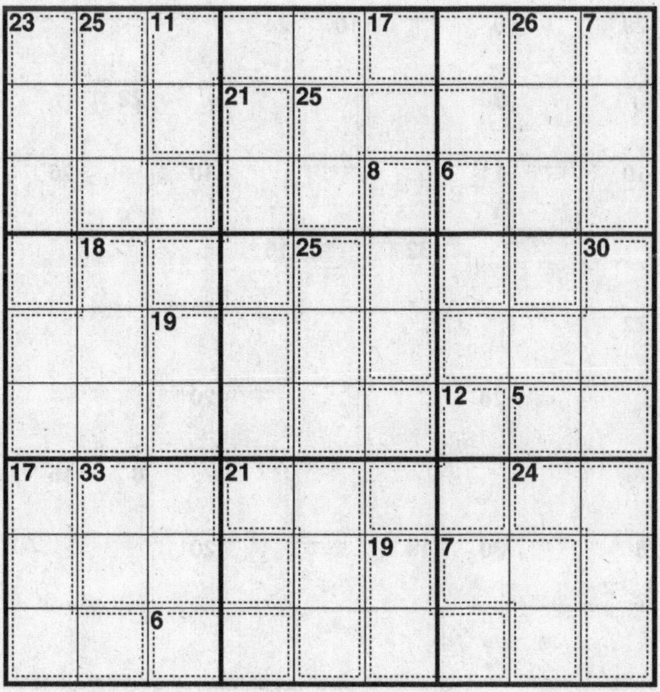

2 HOURS 15 MINUTES

TIME TAKEN..............................

Extra Deadly

112

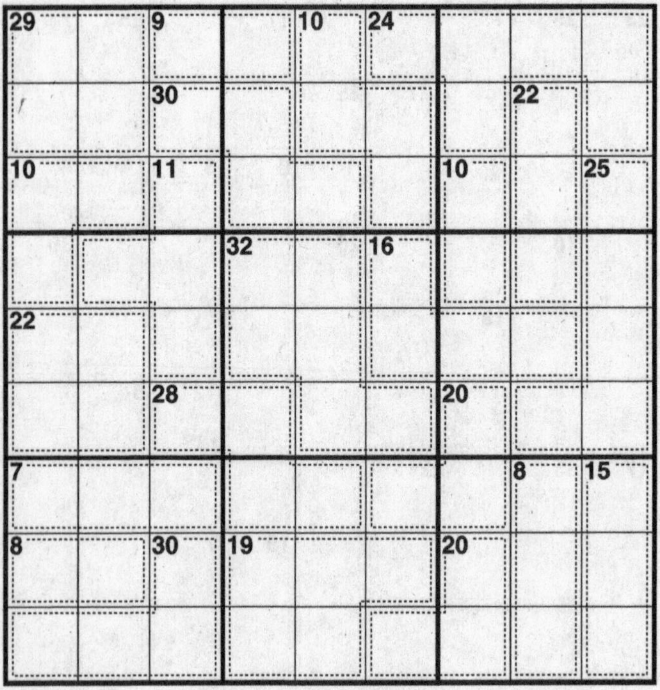

2 HOURS 15 MINUTES

TIME TAKEN..............................

113

2 HOURS 15 MINUTES

TIME TAKEN..............................

Extra Deadly

114

2 HOURS 15 MINUTES

TIME TAKEN..............................

2 HOURS 15 MINUTES

TIME TAKEN..............................

Extra Deadly

116

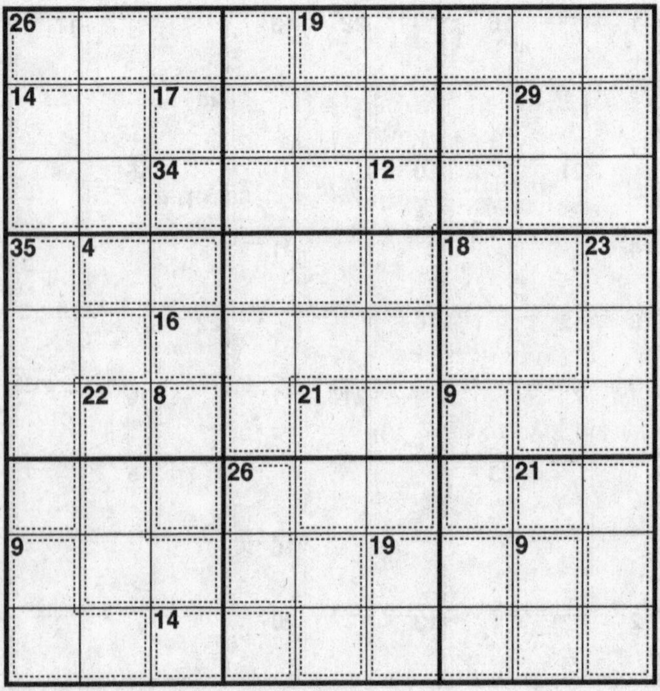

2 HOURS 30 MINUTES

TIME TAKEN..............................

2 HOURS 30 MINUTES

TIME TAKEN..............................

118

2 HOURS 30 MINUTES

TIME TAKEN..............................

119

2 HOURS 30 MINUTES

TIME TAKEN..............................

Extra Deadly

120

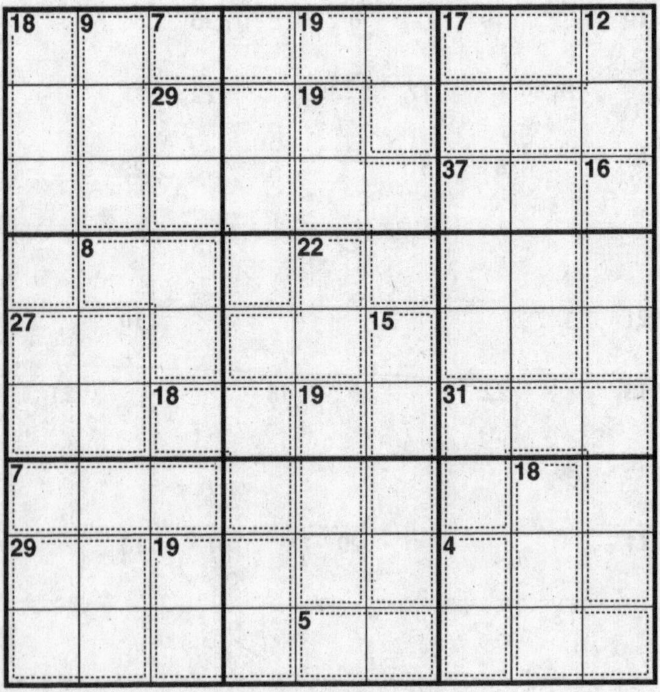

2 HOURS 30 MINUTES

TIME TAKEN..............................

SOLUTIONS

1

[10]1	[17]8	6	3	[13]5	[21]9	7	[29]4	2
9	[7]3	4	7	1	[18]2	5	6	8
[11]2	5	[18]7	6	8	4	3	1	9
3	1	5	[5]4	[28]7	8	9	[13]2	6
[18]6	[13]9	[25]8	1	[5]2	3	4	[18]7	5
7	4	2	[11]9	[11]6	5	[9]1	8	[15]3
5	6	9	2	[11]4	1	8	3	7
[12]4	[10]7	3	[16]8	[12]9	6	[16]2	5	1
8	2	1	5	3	[13]7	6	9	4

2

[19]3	[13]8	5	[18]1	[9]2	4	[12]6	[14]9	[18]7
1	7	2	9	3	[18]6	4	5	8
[31]9	6	[22]4	8	7	5	2	[10]1	3
5	2	9	7	[28]8	[18]1	3	6	[12]4
6	4	3	2	5	9	8	7	1
7	[9]1	8	4	6	[11]3	[11]9	2	[14]5
[7]2	5	[13]7	[15]6	[12]4	8	[9]1	3	9
[24]8	9	6	3	1	7	5	[6]4	2
4	3	1	5	[18]9	2	7	[14]8	6

3

[5]2	3	[8]1	[10]7	[15]9	6	[13]8	[31]4	[19]5
[14]9	5	7	3	[19]8	4	2	6	1
[17]8	[20]4	[8]6	[11]1	5	2	3	7	9
6	7	2	8	[10]3	[6]1	5	9	4
3	9	[12]4	2	7	5	[9]1	8	[11]6
[18]5	1	8	[11]4	6	[19]9	7	3	2
4	8	[16]5	6	1	[29]7	9	2	3
[8]1	2	3	[20]9	4	[15]8	6	5	[16]7
7	[15]6	9	5	2	3	4	1	8

4

[27]9	8	[16]3	5	[16]1	4	[11]6	[9]7	2
6	4	1	7	9	2	5	[9]8	[7]3
[15]7	[7]5	[10]2	8	[25]3	[9]6	[10]9	1	4
8	2	6	9	7	3	1	[6]4	[18]5
[8]3	[13]9	4	[11]6	5	[9]1	8	2	7
5	[25]1	7	4	[10]2	8	[12]3	9	6
2	3	8	[7]1	6	[20]7	4	[14]5	9
[5]1	[13]6	[11]5	2	4	9	[9]7	[9]3	[9]8
4	7	[12]9	3	[13]8	5	2	6	1

5

[11]3	[10]1	9	[12]4	8	[14]5	[18]6	7	[11]2
8	[8]6	[5]4	1	2	7	5	[12]3	9
[35]5	2	7	[9]6	3	[10]9	1	4	[17]8
4	9	8	2	[15]1	6	[11]7	5	3
[7]1	[10]5	3	[19]9	7	8	4	[19]2	6
6	[12]7	2	3	[12]5	4	9	8	[8]1
[21]2	4	1	[11]5	6	3	[10]8	[21]9	7
7	3	[13]5	8	[10]9	1	2	6	[9]4
9	[14]8	6	[11]7	4	2	3	1	5

6

[17]4	8	[11]2	9	[7]6	1	[8]3	5	[26]7
5	[10]1	9	[18]3	8	7	[11]2	6	4
[17]7	6	[7]3	4	[12]5	[17]2	1	8	9
1	3	[11]5	[11]8	7	9	6	[9]4	2
[19]9	[19]4	6	1	2	[20]5	8	7	3
2	7	8	[10]6	3	[13]4	9	[6]1	5
8	[9]5	4	[14]2	1	[11]3	[26]7	9	6
[9]6	[11]2	[8]7	5	[13]9	8	4	[16]3	[9]1
3	9	1	7	4	6	5	2	8

7

[11] 8	[12] 7	5	[19] 4	[15] 9	6	[6] 1	2	3
2	1	[9] 6	7	8	[17] 3	4	9	[18] 5
[16] 4	[14] 9	3	[10] 2	[10] 5	1	[32] 8	6	7
7	4	1	8	2	9	5	3	[10] 6
5	[17] 6	[3] 2	1	3	7	[19] 9	8	4
[10] 9	3	[13] 8	5	[27] 6	4	2	[8] 7	1
1	8	[19] 4	6	7	2	3	[13] 5	[19] 9
[8] 6	2	9	[19] 3	[13] 4	5	7	1	8
[8] 3	5	7	9	1	8	[10] 6	4	2

8

[8] 2	[16] 8	7	[12] 9	[5] 4	1	[7] 3	[19] 5	6
6	[13] 9	1	3	[14] 5	7	4	[10] 2	8
[11] 3	4	[19] 5	8	6	2	7	1	[34] 9
8	[13] 2	[10] 6	1	3	[14] 5	9	4	7
4	7	[25] 3	[25] 6	2	9	5	8	1
[6] 1	5	9	[11] 7	8	[12] 4	2	6	[12] 3
5	[16] 6	8	4	[26] 9	3	[7] 1	7	2
[17] 7	3	2	5	[14] 1	8	6	[20] 9	[9] 4
9	1	4	2	7	6	8	3	5

9

[17] 8	2	[5] 4	[11] 7	[17] 9	3	[12] 6	1	5
7	[20] 3	1	4	5	[34] 6	9	8	[5] 2
[20] 5	9	[13] 6	[10] 2	8	1	[9] 4	7	3
6	8	7	[9] 1	[11] 4	2	5	3	[10] 9
9	[7] 4	[8] 3	8	[25] 7	5	[15] 2	6	1
2	1	5	3	6	9	7	[19] 4	8
[9] 1	5	[20] 9	6	[5] 3	[19] 4	8	[11] 2	7
3	[17] 6	[19] 8	5	2	7	[9] 1	9	[10] 4
4	7	2	9	[9] 1	8	3	5	6

10

[11] 2	9	[16] 6	3	[25] 4	8	[8] 7	1	[30] 5
[13] 1	4	8	5	2	7	[7] 3	6	9
[11] 7	[8] 3	5	[19] 9	[7] 1	6	4	8	2
4	[11] 1	7	8	6	[10] 9	[10] 2	5	3
[31] 8	[13] 5	3	2	[10] 7	1	[7] 6	[13] 9	4
9	6	2	[13] 4	3	[8] 5	1	[15] 7	8
6	[18] 2	9	1	8	3	[26] 5	[7] 4	[8] 7
5	7	[10] 4	6	[11] 9	2	8	3	1
3	[9] 8	1	[12] 7	5	4	9	[8] 2	6

11

[14] 8	2	4	[13] 7	6	[4] 1	[8] 5	3	[11] 9
[10] 7	[15] 9	6	[13] 8	5	3	[5] 4	1	2
3	[6] 1	5	[27] 4	[18] 2	9	7	[11] 8	[10] 6
[14] 5	[15] 7	[13] 3	9	8	6	1	2	4
9	8	1	[22] 3	4	2	[25] 6	[12] 7	5
[10] 4	[9] 6	2	1	7	[20] 5	8	9	[12] 3
6	3	7	5	[10] 9	8	2	[22] 4	1
[12] 2	4	[15] 9	6	1	7	[12] 3	5	8
1	5	[10] 8	2	[7] 3	4	9	6	7

12

[12] 5	[15] 6	9	[26] 7	[16] 1	2	4	[16] 8	[11] 3
7	[6] 3	[12] 8	6	9	[9] 4	[14] 5	2	1
1	2	4	8	[17] 3	5	9	6	7
[13] 3	[12] 4	7	5	6	[9] 1	8	[11] 9	2
2	1	[9] 6	3	8	[20] 9	[13] 7	5	[32] 4
8	[14] 9	5	[18] 2	4	7	[9] 3	1	6
[11] 6	5	[12] 1	9	7	[17] 3	2	4	8
[13] 9	8	3	[11] 4	2	6	[17] 1	7	5
4	[10] 7	2	1	5	8	6	3	9

13

7	2	9	8	3	5	4	6	1
4	3	1	6	7	2	9	8	5
8	5	6	4	1	9	2	7	3
6	4	3	7	9	1	5	2	8
9	1	2	3	5	8	6	4	7
5	8	7	2	4	6	1	3	9
3	6	5	9	2	7	8	1	4
1	7	8	5	6	4	3	9	2
2	9	4	1	8	3	7	5	6

14

8	5	3	1	6	7	4	9	2
9	6	2	5	4	3	7	8	1
7	1	4	8	9	2	5	6	3
1	3	8	2	5	6	9	7	4
5	7	9	3	1	4	6	2	8
2	4	6	7	8	9	1	3	5
4	9	1	6	2	8	3	5	7
6	8	7	4	3	5	2	1	9
3	2	5	9	7	1	8	4	6

15

6	3	9	1	8	4	5	2	7
8	4	5	3	7	2	6	1	9
1	2	7	9	6	5	3	4	8
4	7	8	6	1	9	2	3	5
9	5	6	2	3	7	4	8	1
2	1	3	5	4	8	7	9	6
3	8	2	7	5	1	9	6	4
7	9	4	8	2	6	1	5	3
5	6	1	4	9	3	8	7	2

16

1	2	3	9	7	5	4	8	6
5	6	4	3	2	8	7	9	1
7	9	8	4	6	1	5	2	3
2	1	6	7	9	4	8	3	5
3	5	7	2	8	6	1	4	9
4	8	9	5	1	3	2	6	7
6	7	2	8	5	9	3	1	4
9	4	5	1	3	2	6	7	8
8	3	1	6	4	7	9	5	2

17

5	9	1	6	7	4	8	2	3
3	4	6	5	2	8	7	9	1
2	7	8	3	1	9	5	4	6
4	3	2	8	5	1	9	6	7
7	1	5	9	6	3	2	8	4
8	6	9	2	4	7	1	3	5
1	8	7	4	9	6	3	5	2
6	2	3	1	8	5	4	7	9
9	5	4	7	3	2	6	1	8

18

1	6	8	5	3	7	2	4	9
4	9	5	8	1	2	7	3	6
3	7	2	6	4	9	8	1	5
6	3	1	4	7	8	9	5	2
8	2	9	1	5	6	4	7	3
7	5	4	2	9	3	1	6	8
5	8	6	7	2	4	3	9	1
2	4	3	9	6	1	5	8	7
9	1	7	3	8	5	6	2	4

19

8	6	3	1	9	4	5	7	2
9	5	4	6	7	2	8	1	3
7	2	1	5	8	3	6	9	4
1	9	5	7	3	8	4	2	6
6	7	8	4	2	1	9	3	5
4	3	2	9	5	6	1	8	7
5	1	7	3	6	9	2	4	8
2	4	6	8	1	7	3	5	9
3	8	9	2	4	5	7	6	1

20

4	3	9	7	8	5	6	2	1
5	1	7	3	2	6	8	4	9
8	6	2	4	1	9	5	3	7
3	4	1	8	9	2	7	5	6
2	7	8	5	6	3	1	9	4
9	5	6	1	4	7	3	8	2
1	2	4	6	5	8	9	7	3
6	8	3	9	7	4	2	1	5
7	9	5	2	3	1	4	6	8

21

7	2	4	9	1	8	3	5	6
6	9	5	4	3	2	8	1	7
8	3	1	5	6	7	2	9	4
5	7	2	6	8	9	4	3	1
3	6	8	1	7	4	9	2	5
4	1	9	3	2	5	7	6	8
2	4	3	8	5	6	1	7	9
9	5	7	2	4	1	6	8	3
1	8	6	7	9	3	5	4	2

22

7	5	4	2	3	8	6	1	9
1	9	6	4	5	7	8	3	2
3	8	2	6	9	1	7	5	4
9	1	8	5	4	6	2	7	3
2	3	5	1	7	9	4	6	8
6	4	7	3	8	2	5	9	1
8	2	1	7	6	3	9	4	5
5	7	3	9	2	4	1	8	6
4	6	9	8	1	5	3	2	7

23

4	1	2	6	8	9	3	5	7
7	8	6	1	3	5	9	4	2
5	3	9	4	7	2	8	6	1
6	5	7	9	4	3	1	2	8
3	4	8	7	2	1	5	9	6
2	9	1	8	5	6	7	3	4
9	7	4	5	6	8	2	1	3
8	2	5	3	1	4	6	7	9
1	6	3	2	9	7	4	8	5

24

5	4	3	2	1	7	6	8	9
1	9	2	3	8	6	5	7	4
6	8	7	9	4	5	2	3	1
2	7	6	5	9	3	4	1	8
8	5	1	4	7	2	3	9	6
4	3	9	1	6	8	7	2	5
7	2	4	8	5	9	1	6	3
3	1	8	6	2	4	9	5	7
9	6	5	7	3	1	8	4	2

25

[12] 8	4	[15] 9	6	[9] 7	[17] 3	5	2	[11] 1
[8] 3	5	[11] 1	8	2	[10] 9	7	6	4
[14] 6	7	2	[9] 5	4	1	[12] 9	[13] 8	[11] 3
1	[18] 6	8	4	[14] 9	[15] 7	3	5	2
[14] 2	[19] 9	7	[19] 3	5	8	[9] 1	[20] 4	6
5	3	4	2	1	[12] 6	8	9	7
7	[11] 1	[15] 6	9	[26] 8	4	2	[25] 3	5
[13] 9	2	3	7	6	5	[12] 4	1	8
4	8	5	1	[11] 3	2	6	7	9

26

[7] 2	[10] 4	[19] 3	7	[10] 8	[7] 1	6	[14] 5	9
5	6	9	[24] 4	2	[11] 3	1	7	[17] 8
[16] 1	[15] 7	8	5	9	[25] 6	2	4	3
8	2	[7] 1	6	4	7	[10] 3	[17] 9	[9] 5
[19] 9	5	6	[14] 3	[12] 1	8	7	2	4
7	[8] 3	4	9	5	[10] 2	8	6	[25] 1
3	1	[23] 5	2	6	[18] 4	9	8	7
[27] 4	8	2	1	[10] 7	9	[10] 5	[11] 3	6
6	9	7	8	3	5	4	1	2

27

[19] 4	[6] 2	3	[11] 7	[11] 6	[10] 9	1	[13] 8	5
8	7	1	4	5	[10] 2	[5] 3	[15] 6	9
[14] 5	[19] 9	6	[12] 3	1	8	2	[13] 7	4
9	4	[9] 7	8	[12] 3	[11] 6	5	[13] 1	2
[9] 3	6	2	[11] 1	9	[12] 5	8	4	[16] 7
[6] 1	5	8	2	[14] 4	7	[13] 9	[5] 3	6
[15] 6	[12] 8	[14] 5	9	7	1	4	2	3
7	1	[13] 4	[19] 5	2	[7] 3	[18] 6	[18] 9	8
2	3	9	6	8	4	7	5	1

28

[25] 8	[24] 5	9	7	[9] 1	2	6	[7] 3	4
4	7	6	3	[19] 8	[18] 9	[7] 2	5	[10] 1
[5] 3	[13] 2	[5] 1	4	6	5	[9] 8	[26] 7	9
2	8	3	[11] 9	5	4	1	6	[10] 7
[20] 9	6	[11] 4	2	[10] 7	[9] 1	5	8	3
5	[12] 1	7	[23] 6	3	8	[19] 9	[11] 4	2
[11] 6	9	2	8	4	3	7	[13] 1	5
1	4	[13] 8	5	[12] 2	[13] 7	3	9	[14] 6
[10] 7	3	5	1	9	6	[6] 4	2	8

29

[12] 2	[19] 9	[13] 5	8	[9] 3	6	[12] 4	1	7
3	6	[9] 8	[16] 4	1	[9] 7	2	[22] 9	5
7	4	1	9	2	[20] 5	6	8	[8] 3
[16] 5	2	[10] 6	[15] 7	8	9	[19] 3	4	1
1	8	4	[9] 3	6	2	7	[7] 5	[17] 9
[12] 9	3	[19] 7	5	[24] 4	1	[13] 8	2	6
[19] 4	7	3	1	9	8	5	[13] 6	2
8	1	2	6	[12] 5	3	[17] 9	7	[12] 4
[11] 6	5	[11] 9	2	7	4	1	3	8

30

[18] 3	4	[17] 9	7	1	[8] 2	[12] 5	[19] 6	[12] 8
1	8	2	[12] 4	5	6	7	9	3
[25] 6	[14] 5	[24] 7	[10] 8	3	[19] 9	[8] 2	4	1
8	9	5	2	7	3	6	[5] 1	4
4	3	1	6	[14] 9	5	[19] 8	2	[12] 7
7	2	[18] 6	[9] 1	8	[18] 4	9	[24] 3	5
2	1	4	5	6	8	[11] 3	7	9
[14] 9	[26] 6	8	3	[6] 4	7	1	5	[8] 2
5	[10] 7	3	9	2	[13] 1	4	8	6

31

9	6	4	5	1	8	2	3	7
1	5	3	2	4	7	6	8	9
2	8	7	9	6	3	4	5	1
5	2	8	3	9	1	7	6	4
4	3	9	8	7	6	5	1	2
7	1	6	4	2	5	3	9	8
8	7	5	1	3	4	9	2	6
6	9	1	7	5	2	8	4	3
3	4	2	6	8	9	1	7	5

32

5	1	7	8	2	4	6	3	9
3	6	2	1	9	7	4	8	5
9	8	4	5	3	6	7	1	2
2	4	9	6	8	1	3	5	7
6	7	5	2	4	3	1	9	8
1	3	8	7	5	9	2	4	6
4	5	6	9	1	2	8	7	3
7	9	3	4	6	8	5	2	1
8	2	1	3	7	5	9	6	4

33

8	1	2	7	6	5	3	9	4
4	5	7	2	9	3	6	1	8
3	9	6	4	8	1	7	5	2
6	2	9	8	1	7	4	3	5
7	4	8	3	5	9	1	2	6
1	3	5	6	2	4	9	8	7
5	7	4	1	3	2	8	6	9
9	8	3	5	7	6	2	4	1
2	6	1	9	4	8	5	7	3

34

3	2	8	4	5	9	1	6	7
7	6	5	3	1	2	9	8	4
4	1	9	8	6	7	5	2	3
5	7	2	9	4	3	6	1	8
6	9	4	5	8	1	7	3	2
8	3	1	2	7	6	4	5	9
1	8	3	6	9	4	2	7	5
2	4	7	1	3	5	8	9	6
9	5	6	7	2	8	3	4	1

35

4	6	7	1	9	3	8	5	2
5	3	9	2	6	8	1	4	7
2	8	1	4	7	5	3	9	6
1	5	8	3	2	9	6	7	4
7	9	6	8	1	4	5	2	3
3	4	2	7	5	6	9	8	1
8	1	3	9	4	7	2	6	5
9	7	5	6	3	2	4	1	8
6	2	4	5	8	1	7	3	9

36

2	6	9	3	5	8	7	4	1
5	7	3	2	4	1	8	9	6
8	4	1	6	9	7	2	3	5
4	9	7	1	8	2	6	5	3
3	1	2	9	6	5	4	8	7
6	8	5	4	7	3	9	1	2
7	2	8	5	1	4	3	6	9
1	3	6	8	2	9	5	7	4
9	5	4	7	3	6	1	2	8

37

[17] 1	7	[30] 8	[14] 5	9	[14] 6	[7] 4	3	[9] 2
2	9	5	1	3	4	[14] 6	8	7
3	4	6	2	[15] 8	7	[14] 5	9	[6] 1
[24] 9	8	[13] 1	4	6	[17] 3	[5] 2	[13] 7	5
7	[11] 5	2	[9] 8	1	9	3	6	[12] 4
[15] 4	6	[19] 3	[9] 7	2	5	[10] 9	1	8
8	3	7	[15] 9	[20] 5	2	1	[13] 4	[23] 6
[18] 5	1	9	6	4	8	7	2	3
6	2	4	[10] 3	7	[9] 1	8	5	9

38

[11] 7	4	[11] 8	[6] 1	5	[17] 3	[8] 2	6	[19] 9
[11] 6	[16] 2	3	[11] 8	[10] 4	9	5	[9] 1	7
5	9	1	2	6	[19] 7	[11] 4	8	3
[18] 8	5	[31] 9	6	3	1	7	[6] 2	4
4	7	6	5	[37] 9	2	[7] 1	[12] 3	[14] 8
1	3	2	4	7	8	6	9	5
[18] 3	[12] 8	4	[19] 7	2	6	[14] 9	5	1
9	6	[15] 7	3	[9] 1	5	[15] 8	[11] 4	[8] 2
2	1	5	9	8	4	3	7	6

39

[27] 8	1	3	[14] 9	4	[11] 6	2	[31] 7	5
4	5	6	[12] 7	1	[7] 2	3	9	[18] 8
[25] 9	[9] 7	2	3	[20] 8	5	4	6	1
6	[9] 8	1	2	3	[5] 4	[22] 7	5	9
7	3	[15] 5	6	9	1	8	2	[10] 4
[12] 2	9	4	[13] 8	5	[11] 7	1	3	6
1	[19] 6	9	4	[5] 2	3	[13] 5	8	[12] 7
[12] 5	[17] 2	8	[8] 1	7	[17] 9	[15] 6	[5] 4	3
3	4	7	[11] 5	6	8	9	1	2

40

[22] 9	[6] 2	[14] 8	[9] 1	[18] 4	5	[31] 6	[12] 3	7
7	4	6	8	9	[24] 3	5	2	[11] 1
5	1	[5] 3	2	7	6	8	9	4
[17] 2	9	[5] 1	4	8	[10] 7	3	[20] 5	6
[11] 3	6	[27] 7	9	5	1	[5] 4	8	[11] 2
8	[9] 5	4	6	[10] 3	2	1	7	9
[25] 6	7	9	3	1	[24] 8	[11] 2	4	5
[5] 4	[18] 3	2	[12] 5	6	9	7	[9] 1	8
1	8	5	7	[6] 2	4	[18] 9	6	3

41

[22] 4	8	9	[11] 6	5	[12] 3	1	[7] 2	[26] 7
1	[19] 7	3	9	[6] 2	4	8	5	6
[13] 6	2	5	[13] 8	[13] 7	[8] 1	[8] 4	3	9
[19] 8	9	2	5	6	7	[9] 3	1	4
[12] 7	5	[5] 1	[16] 4	3	[11] 2	6	[26] 9	8
[12] 3	[9] 6	4	[14] 1	8	9	2	7	[11] 5
9	3	6	7	1	[13] 8	5	4	2
[15] 5	1	7	[9] 2	4	[20] 6	[34] 9	8	3
2	[12] 4	8	3	9	5	7	6	1

42

[9] 1	8	[21] 3	5	4	[18] 9	[24] 7	6	2
[21] 5	[12] 6	2	1	8	7	[20] 3	9	[9] 4
9	7	4	[17] 6	[9] 3	2	1	8	5
[12] 8	4	[14] 5	7	6	[6] 1	2	[12] 3	9
[14] 7	1	9	4	2	3	6	[13] 5	[9] 8
2	[9] 3	6	[26] 9	[26] 5	8	4	7	1
4	[11] 2	7	8	9	[23] 6	5	1	[16] 3
[9] 3	9	[14] 1	2	[12] 7	5	8	4	6
6	5	8	[8] 3	1	4	[11] 9	2	7

43

[12] 3	7	[19] 6	8	[6] 2	4	[13] 1	5	[12] 9
[14] 4	2	5	[10] 9	1	[13] 6	[18] 8	7	3
9	[14] 1	8	[8] 5	[13] 3	7	6	[13] 4	2
1	5	[19] 9	3	8	2	4	[17] 6	7
[13] 7	6	2	4	[6] 5	1	[8] 3	9	[9] 8
[12] 8	4	3	[13] 6	7	[14] 9	5	2	1
[8] 2	[11] 3	1	[11] 7	4	5	[32] 9	8	6
6	8	[13] 4	2	[20] 9	3	[13] 7	1	5
[21] 5	9	7	1	6	8	2	3	4

44

[26] 1	9	[17] 7	6	[18] 8	5	3	2	[32] 4
6	2	4	[15] 1	[16] 7	3	5	8	9
[12] 3	8	5	9	4	2	[8] 7	1	6
9	[12] 3	8	[14] 5	[25] 6	[3] 1	2	[13] 4	[12] 7
[17] 7	4	1	3	2	8	[21] 6	9	5
[7] 5	6	2	4	9	7	8	[12] 3	1
2	[8] 1	[19] 3	7	[20] 5	9	[11] 4	6	8
[17] 4	7	9	[20] 8	3	6	1	[25] 5	[5] 2
8	5	6	2	1	4	9	7	3

45

[24] 7	8	[9] 3	[22] 1	2	6	[19] 5	[15] 4	9
9	5	1	[12] 7	4	[11] 3	6	8	2
[6] 4	[15] 6	[7] 2	5	9	8	[11] 3	[8] 1	[10] 7
2	9	5	[16] 6	[11] 1	[18] 4	8	7	3
[17] 6	3	8	2	7	9	[10] 1	5	4
[9] 1	[11] 7	[13] 4	8	3	5	[9] 2	[11] 9	[11] 6
8	4	9	[7] 3	[35] 6	[3] 1	7	2	5
[10] 3	[14] 1	7	4	5	2	[19] 9	6	[12] 8
5	2	6	9	8	7	4	3	1

46

[9] 1	[16] 3	4	9	[19] 8	6	[8] 2	[31] 7	[9] 5
8	[9] 2	7	[30] 1	[11] 4	5	6	9	3
[18] 9	5	6	2	7	3	4	8	1
2	9	3	4	[19] 6	1	7	[13] 5	8
7	[18] 8	[12] 1	[13] 5	3	2	[18] 9	[9] 6	[14] 4
4	6	5	8	[16] 9	7	1	3	2
[14] 3	4	2	[20] 6	5	9	8	1	7
6	[8] 7	[17] 8	[13] 3	1	[11] 4	5	2	[19] 9
5	1	9	7	2	[11] 8	3	4	6

47

[9] 3	1	[22] 9	8	[9] 2	[9] 5	[19] 4	6	[8] 7
5	[13] 6	2	3	7	4	[19] 8	9	1
[13] 4	7	[15] 8	[15] 9	6	[11] 1	2	3	5
9	[13] 8	7	[18] 5	[12] 3	2	1	[24] 4	[8] 6
1	4	6	7	9	8	[19] 3	5	2
[10] 2	3	5	[11] 1	[10] 4	6	9	7	8
[26] 6	5	[19] 4	2	8	[30] 3	7	[10] 1	9
7	2	1	[9] 4	5	9	6	[11] 8	3
8	9	3	[7] 6	1	7	5	[6] 2	4

48

[13] 6	7	[15] 5	1	9	[13] 2	8	3	[31] 4
[20] 8	2	1	[16] 3	[17] 7	4	9	5	6
9	[14] 4	3	8	5	6	7	[12] 2	1
[18] 4	5	2	[12] 6	3	[9] 8	1	9	[10] 7
1	8	[11] 7	4	2	[20] 9	5	6	3
3	[15] 9	6	[26] 5	1	[8] 7	[12] 2	4	[20] 8
2	[10] 3	4	9	8	1	6	7	5
[12] 5	1	[19] 9	[13] 7	6	[7] 3	4	[12] 8	[11] 2
7	6	8	2	[9] 4	5	3	1	9

49

9	5	2	4	7	3	6	1	8
6	7	8	5	2	1	3	4	9
4	3	1	9	6	8	5	7	2
2	4	6	8	1	5	9	3	7
5	9	3	2	4	7	8	6	1
1	8	7	6	3	9	4	2	5
7	6	9	1	5	4	2	8	3
3	2	5	7	8	6	1	9	4
8	1	4	3	9	2	7	5	6

50

4	7	5	6	1	9	2	8	3
6	1	8	3	5	2	4	9	7
9	3	2	4	8	7	5	1	6
2	8	4	1	6	5	7	3	9
1	9	7	2	3	4	6	5	8
3	5	6	9	7	8	1	4	2
5	6	3	7	9	1	8	2	4
7	4	1	8	2	3	9	6	5
8	2	9	5	4	6	3	7	1

51

8	2	4	1	9	3	7	6	5
6	3	7	2	4	5	9	1	8
1	5	9	8	7	6	3	2	4
4	1	5	3	8	2	6	7	9
9	8	3	5	6	7	1	4	2
2	7	6	4	1	9	5	8	3
5	6	2	7	3	8	4	9	1
3	9	1	6	2	4	8	5	7
7	4	8	9	5	1	2	3	6

52

3	4	7	9	1	2	8	5	6
6	1	9	7	8	5	2	3	4
5	2	8	6	4	3	9	1	7
7	9	6	4	3	8	1	2	5
2	3	5	1	7	6	4	9	8
4	8	1	5	2	9	6	7	3
9	5	4	2	6	7	3	8	1
8	6	2	3	5	1	7	4	9
1	7	3	8	9	4	5	6	2

53

9	3	5	1	2	7	4	6	8
8	4	1	3	9	6	2	5	7
6	7	2	8	4	5	1	3	9
2	9	4	7	6	8	3	1	5
5	1	8	2	3	9	6	7	4
7	6	3	5	1	4	8	9	2
3	2	9	4	5	1	7	8	6
4	8	6	9	7	3	5	2	1
1	5	7	6	8	2	9	4	3

54

7	1	6	2	8	4	9	3	5
9	2	5	3	1	6	4	7	8
3	4	8	7	5	9	2	6	1
8	5	9	1	3	2	7	4	6
2	7	3	4	6	8	5	1	9
1	6	4	9	7	5	3	8	2
5	8	7	6	2	3	1	9	4
4	3	2	8	9	1	6	5	7
6	9	1	5	4	7	8	2	3

55

3	8	4	6	9	1	5	2	7
1	9	7	3	2	5	4	6	8
2	5	6	7	4	8	9	3	1
9	2	1	5	8	4	6	7	3
7	3	5	9	1	6	8	4	2
6	4	8	2	7	3	1	5	9
8	7	2	4	6	9	3	1	5
4	1	3	8	5	7	2	9	6
5	6	9	1	3	2	7	8	4

56

6	9	1	5	8	2	4	3	7
8	7	5	3	6	4	9	2	1
3	4	2	7	1	9	6	5	8
7	2	6	1	3	8	5	4	9
9	5	3	2	4	7	8	1	6
1	8	4	6	9	5	2	7	3
4	3	7	8	5	6	1	9	2
5	1	8	9	2	3	7	6	4
2	6	9	4	7	1	3	8	5

57

4	7	1	9	8	2	5	6	3
5	8	9	1	6	3	4	7	2
2	3	6	7	5	4	1	9	8
1	6	3	2	7	9	8	4	5
7	4	5	3	1	8	9	2	6
9	2	8	5	4	6	3	1	7
8	1	7	4	2	5	6	3	9
3	5	2	6	9	1	7	8	4
6	9	4	8	3	7	2	5	1

58

2	8	1	6	4	7	9	5	3
4	9	7	5	2	3	8	6	1
5	6	3	9	1	8	2	4	7
7	5	8	1	3	6	4	9	2
6	2	4	8	7	9	1	3	5
3	1	9	4	5	2	6	7	8
1	7	6	2	9	5	3	8	4
9	4	5	3	8	1	7	2	6
8	3	2	7	6	4	5	1	9

59

7	8	9	4	6	2	3	1	5
6	3	1	8	5	7	4	9	2
2	4	5	1	3	9	8	7	6
8	9	2	3	1	5	6	4	7
5	1	4	2	7	6	9	3	8
3	6	7	9	8	4	5	2	1
1	5	6	7	9	3	2	8	4
4	7	3	5	2	8	1	6	9
9	2	8	6	4	1	7	5	3

60

8	9	2	6	7	3	5	1	4
4	7	3	2	1	5	9	6	8
1	5	6	8	9	4	3	2	7
5	1	8	4	2	7	6	9	3
2	6	4	5	3	9	8	7	1
9	3	7	1	6	8	4	5	2
7	2	9	3	8	6	1	4	5
3	4	1	9	5	2	7	8	6
6	8	5	7	4	1	2	3	9

61

[19]9	4	[14]1	6	[19]7	[29]5	8	[14]3	2
6	2	5	9	3	[14]8	7	4	[7]1
[10]7	[9]3	[9]8	1	2	4	9	5	6
3	6	[20]9	[15]7	[7]1	[11]2	[17]4	8	5
[6]5	7	4	8	6	9	[11]2	[10]1	[10]3
1	[19]8	[11]2	[9]4	5	3	6	9	7
2	9	6	3	[18]8	1	[10]5	[35]7	4
[17]4	[11]1	7	[11]5	9	[13]6	3	2	8
8	5	3	2	4	7	1	6	9

62

[20]5	[12]9	3	[20]8	6	1	2	[11]4	7
7	[13]4	[17]6	[18]9	2	3	[9]8	1	[21]5
8	1	2	[12]4	7	[11]5	[18]3	6	9
[11]9	8	5	7	4	2	6	[8]3	1
2	[14]7	4	1	[9]3	6	9	5	[14]8
[9]3	6	1	[13]5	8	[20]9	4	7	2
6	[26]2	7	[17]3	5	[19]8	[13]1	[11]9	4
[13]1	5	8	[9]6	9	4	7	2	[17]3
4	3	9	2	1	7	5	8	6

63

[8]7	1	[5]2	[15]5	[10]6	[10]3	[5]4	[19]8	9
[26]9	5	3	8	4	7	1	[13]6	2
8	[13]6	4	2	[15]1	9	5	7	[13]3
4	3	[26]6	[24]7	9	5	[10]8	2	1
[9]1	8	7	[7]6	3	[11]2	9	5	4
[11]2	9	5	1	[19]8	[14]4	7	[16]3	[18]6
[14]6	[14]7	8	9	2	[22]1	3	4	5
3	2	1	[9]4	5	8	[17]6	9	7
5	4	[12]9	3	7	6	2	1	8

64

[14]4	[7]5	[18]7	2	[6]1	[11]3	[20]9	8	[17]6
1	2	9	[25]6	5	8	3	7	4
3	6	8	7	[12]9	[12]4	2	[14]1	5
[12]2	1	4	[13]9	3	6	[12]7	5	8
9	[13]8	5	4	[8]7	1	[15]6	3	2
[21]6	7	[5]3	[12]5	[15]8	2	[5]1	4	[17]9
8	[12]3	2	1	6	5	4	[30]9	7
[12]5	9	[9]6	3	4	7	8	2	1
7	[15]4	1	8	2	[14]9	5	[9]6	3

65

[22]3	4	7	[14]6	8	[12]2	9	1	[21]5
[13]5	8	[10]9	[8]4	[10]1	7	2	3	6
2	[22]6	1	3	[14]9	5	[12]4	8	7
6	9	[13]8	1	[28]4	[15]3	5	7	[5]2
[5]1	7	5	2	6	9	[9]8	[19]4	3
4	[10]3	2	5	7	[13]8	1	6	[21]9
[25]7	[8]2	6	[13]8	5	1	[9]3	9	4
9	1	[12]3	7	2	4	6	[15]5	8
8	[18]5	4	9	[9]3	6	7	2	1

66

[19]6	8	5	[16]4	[19]9	3	[17]1	7	[5]2
[6]4	2	1	5	6	7	9	[13]8	3
[12]9	[21]7	[17]3	[9]8	1	[6]2	[14]6	5	[13]4
3	5	8	6	[17]7	4	2	1	9
2	6	[16]4	9	8	[6]1	5	[10]3	7
1	[19]9	7	3	2	5	[17]4	6	[13]8
[20]8	3	[19]6	1	[19]4	9	7	[19]2	5
5	[6]4	2	7	3	6	8	9	[11]1
7	[10]1	9	2	[16]5	8	3	4	6

67

6	7	3	2	1	9	5	8	4
1	4	2	6	8	5	9	7	3
5	8	9	3	7	4	1	2	6
3	5	8	4	2	7	6	1	9
4	1	7	9	6	8	2	3	5
2	9	6	1	5	3	8	4	7
9	3	1	8	4	6	7	5	2
7	2	4	5	9	1	3	6	8
8	6	5	7	3	2	4	9	1

68

6	9	4	2	7	1	5	3	8
2	1	7	8	3	5	6	4	9
8	3	5	6	4	9	7	1	2
7	2	3	4	1	6	9	8	5
4	5	8	7	9	3	2	6	1
9	6	1	5	8	2	3	7	4
1	7	9	3	5	4	8	2	6
5	8	2	1	6	7	4	9	3
3	4	6	9	2	8	1	5	7

69

7	2	5	3	4	8	9	6	1
8	6	4	1	7	9	5	3	2
9	3	1	5	2	6	7	8	4
5	8	9	2	6	7	1	4	3
3	7	6	4	5	1	2	9	8
4	1	2	8	9	3	6	5	7
6	4	8	9	1	2	3	7	5
1	9	3	7	8	5	4	2	6
2	5	7	6	3	4	8	1	9

70

9	2	5	1	8	6	3	4	7
4	6	8	5	3	7	9	1	2
3	1	7	2	9	4	5	8	6
2	5	6	8	1	3	7	9	4
1	3	9	4	7	2	8	6	5
7	8	4	6	5	9	2	3	1
8	7	1	9	6	5	4	2	3
6	4	3	7	2	8	1	5	9
5	9	2	3	4	1	6	7	8

71

7	4	9	5	1	6	2	8	3
5	8	3	2	4	9	6	1	7
1	6	2	3	8	7	5	4	9
4	1	8	7	6	2	3	9	5
6	2	5	9	3	4	1	7	8
3	9	7	1	5	8	4	6	2
2	7	6	4	9	3	8	5	1
9	5	4	8	2	1	7	3	6
8	3	1	6	7	5	9	2	4

72

2	7	8	4	6	5	1	3	9
1	6	4	9	3	2	8	7	5
5	9	3	1	8	7	4	6	2
3	5	7	8	2	9	6	4	1
4	8	6	5	1	3	2	9	7
9	2	1	6	7	4	5	8	3
6	4	5	7	9	1	3	2	8
7	1	2	3	4	8	9	5	6
8	3	9	2	5	6	7	1	4

73

2	3	6	8	1	4	9	5	7
4	5	9	2	3	7	8	6	1
7	1	8	5	6	9	2	3	4
1	9	7	4	5	6	3	2	8
8	2	4	9	7	3	6	1	5
3	6	5	1	8	2	7	4	9
5	8	3	7	2	1	4	9	6
9	7	2	6	4	5	1	8	3
6	4	1	3	9	8	5	7	2

74

2	4	8	5	9	1	3	6	7
6	7	9	8	3	4	5	1	2
5	3	1	2	7	6	8	4	9
3	6	4	1	2	9	7	8	5
7	8	2	6	4	5	9	3	1
9	1	5	7	8	3	6	2	4
1	2	7	9	6	8	4	5	3
8	5	3	4	1	7	2	9	6
4	9	6	3	5	2	1	7	8

75

3	4	2	7	9	1	5	6	8
8	5	6	2	4	3	7	1	9
1	9	7	8	5	6	3	2	4
2	1	3	6	8	4	9	7	5
7	8	4	5	2	9	6	3	1
5	6	9	1	3	7	4	8	2
9	7	1	4	6	8	2	5	3
6	3	5	9	1	2	8	4	7
4	2	8	3	7	5	1	9	6

76

5	9	2	7	6	4	8	3	1
3	7	6	8	9	1	4	2	5
4	8	1	3	5	2	7	6	9
9	3	7	2	1	5	6	4	8
1	5	4	6	3	8	9	7	2
6	2	8	4	7	9	1	5	3
7	1	9	5	2	6	3	8	4
2	4	3	9	8	7	5	1	6
8	6	5	1	4	3	2	9	7

77

4	9	8	1	6	7	2	3	5
2	7	5	4	3	9	1	6	8
3	6	1	5	2	8	4	7	9
7	4	9	6	1	5	8	2	3
5	8	6	2	9	3	7	4	1
1	2	3	7	8	4	5	9	6
9	1	7	3	5	2	6	8	4
8	5	4	9	7	6	3	1	2
6	3	2	8	4	1	9	5	7

78

1	6	8	5	9	3	7	4	2
4	9	7	8	6	2	5	3	1
2	5	3	1	4	7	8	6	9
7	8	4	2	5	6	9	1	3
9	2	1	3	7	4	6	5	8
5	3	6	9	1	8	4	2	7
3	1	5	6	8	9	2	7	4
8	4	2	7	3	5	1	9	6
6	7	9	4	2	1	3	8	5

79

[15]4	3	6	[12]1	7	[24]2	5	8	9
[13]5	2	[16]7	4	[22]9	8	[10]1	6	3
8	[18]9	1	[13]6	5	[10]3	7	[20]4	[10]2
[24]2	1	3	5	[10]6	4	9	7	8
7	8	5	2	[12]3	9	[11]4	1	6
9	6	[13]4	[15]8	[15]1	[9]7	2	[8]3	5
[14]6	4	9	7	8	[10]5	3	2	[12]1
3	[20]7	8	[14]9	2	[7]1	6	[14]5	4
1	5	2	3	4	[14]6	8	9	7

80

[21]6	7	1	[14]2	3	[9]4	5	[30]9	8
4	3	[22]8	[13]1	9	[7]5	2	[9]6	7
[7]2	5	9	8	[26]6	7	1	3	4
[12]8	[10]6	5	4	[17]7	9	3	[6]1	2
1	4	[9]2	[15]3	8	[13]6	7	5	[15]9
3	[10]9	7	5	2	[11]1	4	[13]8	6
[21]9	1	[19]4	7	[13]5	8	6	2	3
5	[13]2	6	9	[7]4	3	[30]8	7	[6]1
7	8	3	[7]6	1	2	9	4	5

81

[4]1	3	[14]2	5	7	[11]8	[26]9	[11]4	[8]6
[10]6	4	[14]5	[10]9	1	3	8	7	2
[19]7	8	9	[17]4	2	6	3	[13]5	1
4	[26]1	6	8	3	[16]9	5	2	7
9	7	3	[14]1	5	2	6	[12]8	4
[8]5	[10]2	8	[10]7	[25]6	4	[5]1	[15]9	[12]3
3	[31]5	1	2	8	7	4	6	9
[10]2	6	4	[9]3	[14]9	5	[9]7	[17]1	8
8	9	7	6	[5]4	1	2	3	5

82

[26]9	8	4	[7]3	[6]5	1	[9]2	7	[26]6
5	[9]3	6	4	[21]7	[11]2	[5]1	8	9
[13]2	1	[17]7	6	8	9	4	3	[17]5
4	6	9	[15]7	[14]1	[8]5	3	[7]2	8
[14]7	2	1	8	6	3	[15]9	5	4
[11]3	5	[10]8	[7]2	[12]9	4	6	[10]1	[10]7
8	[11]4	2	5	3	[26]6	7	9	1
[16]1	7	[24]3	9	4	8	5	[13]6	2
6	9	5	1	2	[15]7	8	4	3

83

[19]8	[11]7	3	1	[33]4	6	9	5	2
5	[11]6	1	[11]3	[14]9	[9]2	[14]4	7	[16]8
2	4	[15]9	8	5	7	3	[15]6	1
1	[19]2	6	[15]4	[18]8	3	5	9	7
3	9	8	5	7	[12]1	2	[13]4	6
[19]4	[12]5	7	6	2	9	[9]8	1	3
7	8	[6]2	[15]9	6	[13]4	1	3	[35]5
[19]6	1	4	[5]2	3	5	7	8	9
9	3	[13]5	7	1	[14]8	6	2	4

84

[20]5	8	7	[26]9	6	1	2	[14]4	3
[10]6	4	[18]9	[12]8	[9]2	3	[6]5	1	7
[6]2	[10]3	1	4	7	5	[14]8	6	[10]9
4	7	8	[14]6	3	[11]2	[20]9	5	1
[12]3	9	[16]2	5	1	8	6	[11]7	4
[9]1	6	5	[17]7	9	[10]4	[10]3	[20]8	[7]2
8	[24]2	3	1	[11]4	6	7	9	5
9	1	[10]6	2	5	[11]7	4	3	[14]8
7	5	4	[11]3	8	[12]9	1	2	6

85

8	1	4	5	7	2	9	3	6
5	7	6	4	9	3	2	8	1
9	3	2	6	1	8	7	5	4
3	5	9	1	8	6	4	7	2
4	8	7	3	2	9	6	1	5
6	2	1	7	5	4	3	9	8
2	9	3	8	6	5	1	4	7
7	6	8	9	4	1	5	2	3
1	4	5	2	3	7	8	6	9

86

8	5	2	3	4	7	9	1	6
7	1	3	9	5	6	4	8	2
6	4	9	8	1	2	5	7	3
3	9	8	7	6	5	1	2	4
2	7	5	4	8	1	3	6	9
1	6	4	2	3	9	7	5	8
5	2	7	6	9	4	8	3	1
4	3	1	5	2	8	6	9	7
9	8	6	1	7	3	2	4	5

87

9	7	3	5	1	2	8	4	6
4	2	8	6	3	7	9	5	1
5	6	1	9	8	4	2	3	7
6	5	7	3	2	9	1	8	4
8	4	9	7	6	1	3	2	5
1	3	2	4	5	8	6	7	9
3	9	4	8	7	6	5	1	2
2	8	6	1	4	5	7	9	3
7	1	5	2	9	3	4	6	8

88

2	1	8	6	3	7	5	4	9
6	9	4	8	5	2	7	3	1
7	3	5	4	9	1	2	8	6
3	7	2	1	8	9	6	5	4
5	4	1	3	7	6	9	2	8
8	6	9	5	2	4	3	1	7
1	5	3	9	6	8	4	7	2
4	2	6	7	1	5	8	9	3
9	8	7	2	4	3	1	6	5

89

4	7	1	3	6	8	9	2	5
9	3	2	4	7	5	8	6	1
5	8	6	9	1	2	7	3	4
2	6	7	8	9	1	4	5	3
1	4	5	6	3	7	2	8	9
8	9	3	2	5	4	1	7	6
3	2	4	5	8	9	6	1	7
7	5	8	1	4	6	3	9	2
6	1	9	7	2	3	5	4	8

90

9	5	8	7	4	1	6	3	2
6	2	1	9	3	5	7	8	4
7	4	3	8	2	6	9	5	1
2	8	4	5	9	3	1	6	7
1	6	9	4	7	8	3	2	5
5	3	7	6	1	2	4	9	8
4	7	6	2	5	9	8	1	3
3	9	5	1	8	7	2	4	6
8	1	2	3	6	4	5	7	9

91

9	6	5	3	8	2	1	4	7
3	4	1	7	6	5	2	9	8
8	7	2	1	4	9	5	6	3
4	1	3	5	9	8	7	2	6
7	2	6	4	1	3	8	5	9
5	8	9	6	2	7	4	3	1
1	5	4	8	3	6	9	7	2
6	9	8	2	7	4	3	1	5
2	3	7	9	5	1	6	8	4

92

8	2	4	5	7	1	6	3	9
3	1	5	4	6	9	7	8	2
9	6	7	2	3	8	1	5	4
5	4	3	8	2	7	9	1	6
6	9	8	1	4	5	2	7	3
2	7	1	6	9	3	8	4	5
1	5	2	3	8	6	4	9	7
7	3	6	9	1	4	5	2	8
4	8	9	7	5	2	3	6	1

93

5	1	9	3	6	2	7	8	4
4	2	6	7	8	5	1	9	3
8	7	3	1	9	4	6	5	2
3	6	1	5	7	9	2	4	8
7	4	2	8	3	1	5	6	9
9	8	5	4	2	6	3	1	7
1	3	4	2	5	8	9	7	6
6	5	7	9	4	3	8	2	1
2	9	8	6	1	7	4	3	5

94

7	3	6	8	9	5	2	1	4
5	1	8	3	4	2	9	7	6
9	2	4	6	7	1	5	8	3
4	9	2	5	1	8	6	3	7
3	7	1	4	6	9	8	5	2
6	8	5	2	3	7	4	9	1
2	5	7	1	8	6	3	4	9
1	6	3	9	5	4	7	2	8
8	4	9	7	2	3	1	6	5

95

9	8	5	4	1	6	2	3	7
3	6	1	8	7	2	4	9	5
2	4	7	5	9	3	1	8	6
1	9	8	6	4	7	3	5	2
6	5	2	3	8	1	9	7	4
7	3	4	2	5	9	8	6	1
4	2	3	9	6	5	7	1	8
5	7	9	1	2	8	6	4	3
8	1	6	7	3	4	5	2	9

96

4	9	8	3	7	5	2	1	6
7	2	6	9	1	4	8	3	5
5	1	3	8	2	6	4	9	7
2	4	9	6	5	1	7	8	3
3	6	5	7	9	8	1	4	2
1	8	7	4	3	2	5	6	9
8	3	2	5	4	9	6	7	1
6	7	1	2	8	3	9	5	4
9	5	4	1	6	7	3	2	8

97

[18]9	[9]2	[18]3	7	[11]4	1	[20]5	[14]6	8
4	7	8	[10]2	6	[25]5	1	3	[16]9
5	[17]1	[14]6	8	3	9	4	7	2
2	6	4	3	[18]7	8	[10]9	1	5
3	5	[17]7	1	9	[25]4	8	[6]2	[9]6
[9]8	9	1	[11]5	2	6	7	4	3
1	[11]8	2	4	[8]5	3	[15]6	9	[11]7
[26]6	3	[14]5	9	[8]1	7	[19]2	8	4
7	4	9	[16]6	8	2	3	5	1

98

[12]8	[17]2	6	[13]1	5	7	[16]3	4	9
4	9	[10]7	3	[19]2	[11]6	[9]1	[19]8	5
[12]3	5	[36]1	8	9	4	2	[17]7	6
[9]2	4	[8]3	5	8	1	6	9	[10]7
7	[9]8	5	9	6	[9]2	4	1	3
[11]6	1	[19]9	[11]4	7	3	[16]8	5	2
5	6	4	7	[7]3	[17]8	[14]9	2	1
[8]1	7	[10]2	[14]6	4	9	[12]5	3	[18]8
[12]9	3	8	2	1	5	7	6	4

99

[12]5	7	[8]2	[14]6	4	3	1	[28]8	9
[9]1	8	6	[18]2	[14]5	9	4	7	[10]3
[7]4	3	9	7	[18]8	[7]1	6	5	2
[25]8	[9]5	4	[10]1	3	[15]6	[19]9	2	[8]7
6	2	[11]3	9	7	5	8	[7]4	1
9	[26]1	7	[11]8	[19]2	4	[7]5	3	[23]6
7	4	1	3	9	8	2	[15]6	5
3	6	5	[5]4	1	[9]2	[11]7	9	8
[19]2	9	8	[11]5	6	7	3	1	4

100

[12]7	5	[7]4	[14]2	3	1	[19]9	[13]6	[13]8
[8]2	[31]1	3	[21]9	8	6	4	7	5
6	9	8	7	[11]4	5	2	[18]1	[13]3
[16]8	6	7	5	[14]1	[16]4	3	2	9
5	[6]2	[12]9	6	7	3	8	4	1
3	4	1	[12]8	[18]2	9	[15]6	[18]5	7
[9]1	8	2	4	9	7	5	3	6
[20]4	7	[8]5	3	[12]6	[10]8	1	[11]9	2
9	[9]3	6	1	5	2	[19]7	8	4

101

[11]5	6	[10]7	[6]1	[12]9	3	[13]8	4	[9]2
[12]4	8	3	5	[8]2	6	[14]9	1	7
[10]1	[11]9	2	[19]8	7	[15]4	5	[9]3	6
9	[15]7	[12]4	3	1	5	6	[11]2	8
2	5	8	[8]6	[13]4	9	[24]3	7	1
[9]3	1	[18]6	2	[15]8	7	[15]4	5	9
6	2	5	[24]7	3	8	1	[19]9	4
[31]8	4	1	9	5	2	[16]7	6	[16]3
7	3	9	4	6	1	2	8	5

102

[20]7	[12]2	[13]9	[12]4	8	[15]1	5	[9]3	6
8	4	1	[13]3	[31]5	6	9	[17]2	7
5	6	3	2	[9]7	9	[11]4	8	[12]1
[12]3	[17]5	4	8	2	7	6	1	9
9	8	[26]6	[7]1	[13]3	4	[16]7	5	2
[9]2	1	7	6	9	[8]5	3	4	[12]8
6	[10]7	8	5	1	[11]3	[11]2	9	4
[14]4	3	[33]2	9	6	8	[11]1	[13]7	[8]5
1	9	5	7	4	2	8	6	3

103

8	4	7	5	6	1	9	2	3
2	3	6	4	7	9	8	1	5
1	5	9	2	3	8	7	6	4
4	8	5	7	2	6	1	3	9
7	2	3	9	1	4	5	8	6
9	6	1	3	8	5	2	4	7
6	7	2	1	5	3	4	9	8
5	9	8	6	4	2	3	7	1
3	1	4	8	9	7	6	5	2

104

5	1	2	8	6	3	7	4	9
6	4	3	1	9	7	8	2	5
7	9	8	5	2	4	6	3	1
9	8	4	2	7	6	1	5	3
1	5	6	3	8	9	4	7	2
2	3	7	4	5	1	9	8	6
4	2	9	7	1	5	3	6	8
8	7	1	6	3	2	5	9	4
3	6	5	9	4	8	2	1	7

105

5	7	6	8	9	4	3	1	2
2	4	8	3	7	1	5	6	9
1	9	3	2	6	5	8	4	7
4	2	9	5	3	8	6	7	1
7	6	5	1	4	9	2	8	3
8	3	1	6	2	7	9	5	4
9	5	4	7	8	3	1	2	6
3	1	2	4	5	6	7	9	8
6	8	7	9	1	2	4	3	5

106

3	6	5	8	4	2	1	9	7
8	9	7	6	5	1	4	2	3
1	4	2	3	7	9	6	8	5
5	7	1	9	8	4	3	6	2
2	3	6	5	1	7	9	4	8
4	8	9	2	3	6	5	7	1
7	5	4	1	9	8	2	3	6
9	2	3	7	6	5	8	1	4
6	1	8	4	2	3	7	5	9

107

6	8	9	2	7	3	4	5	1
2	5	1	6	4	8	3	9	7
4	3	7	1	9	5	2	8	6
1	2	5	7	8	6	9	3	4
3	6	8	9	5	4	7	1	2
9	7	4	3	2	1	5	6	8
7	9	3	8	6	2	1	4	5
5	1	6	4	3	7	8	2	9
8	4	2	5	1	9	6	7	3

108

7	3	2	5	6	8	9	1	4
8	1	5	4	3	9	7	6	2
9	4	6	2	7	1	5	3	8
5	7	9	6	1	4	8	2	3
3	6	8	7	9	2	4	5	1
1	2	4	8	5	3	6	9	7
6	5	3	1	4	7	2	8	9
2	9	7	3	8	5	1	4	6
4	8	1	9	2	6	3	7	5

109

9	6	2	3	7	5	4	8	1
3	8	4	6	2	1	5	7	9
7	5	1	4	9	8	2	3	6
2	1	9	8	4	7	3	6	5
4	3	8	5	1	6	7	9	2
6	7	5	9	3	2	8	1	4
5	2	7	1	8	9	6	4	3
1	4	6	7	5	3	9	2	8
8	9	3	2	6	4	1	5	7

110

3	4	2	7	1	5	9	8	6
1	7	6	3	8	9	5	4	2
8	9	5	2	4	6	3	7	1
5	8	9	4	3	2	6	1	7
7	3	4	6	5	1	8	2	9
6	2	1	8	9	7	4	5	3
9	5	7	1	6	4	2	3	8
4	1	8	9	2	3	7	6	5
2	6	3	5	7	8	1	9	4

111

3	5	6	9	4	7	1	2	8
7	9	1	2	6	8	4	5	3
4	2	8	1	5	3	7	6	9
5	6	7	4	8	1	9	3	2
2	8	3	6	7	9	5	1	4
1	4	9	3	2	5	8	7	6
6	7	4	5	9	2	3	8	1
8	1	2	7	3	4	6	9	5
9	3	5	8	1	6	2	4	7

112

7	3	6	8	5	9	4	1	2
2	1	5	4	3	6	7	9	8
4	8	9	7	1	2	5	6	3
9	7	1	5	6	8	3	2	4
3	4	2	9	7	1	8	5	6
6	5	8	2	4	3	9	7	1
1	6	7	3	9	4	2	8	5
5	2	3	6	8	7	1	4	9
8	9	4	1	2	5	6	3	7

113

6	7	1	4	2	9	3	8	5
4	8	2	3	7	5	9	6	1
9	3	5	6	8	1	7	4	2
5	6	9	8	1	7	4	2	3
1	2	8	5	3	4	6	7	9
7	4	3	9	6	2	5	1	8
8	5	7	2	4	3	1	9	6
3	1	6	7	9	8	2	5	4
2	9	4	1	5	6	8	3	7

114

4	9	8	7	5	6	1	2	3
7	6	2	1	9	3	5	4	8
3	5	1	2	4	8	7	6	9
8	1	7	6	3	5	4	9	2
5	4	9	8	7	2	3	1	6
6	2	3	9	1	4	8	5	7
2	8	4	3	6	1	9	7	5
1	7	6	5	8	9	2	3	4
9	3	5	4	2	7	6	8	1

115

9	5	7	8	3	6	1	4	2
2	1	4	7	5	9	8	6	3
3	8	6	2	1	4	9	7	5
1	2	3	4	6	7	5	9	8
7	9	5	1	8	2	6	3	4
6	4	8	3	9	5	2	1	7
8	6	9	5	7	3	4	2	1
4	7	1	9	2	8	3	5	6
5	3	2	6	4	1	7	8	9

116

7	2	8	3	9	4	1	6	5
3	4	1	8	6	5	9	2	7
6	9	5	1	2	7	3	8	4
9	8	3	6	5	2	7	4	1
2	5	7	9	4	1	6	3	8
1	6	4	7	8	3	5	9	2
5	7	2	4	3	6	8	1	9
4	3	9	5	1	8	2	7	6
8	1	6	2	7	9	4	5	3

117

2	1	7	4	8	5	9	6	3
3	8	5	9	7	6	4	1	2
6	9	4	2	1	3	8	5	7
8	7	1	5	2	4	6	3	9
9	5	3	1	6	7	2	8	4
4	2	6	3	9	8	5	7	1
5	3	8	7	4	9	1	2	6
7	4	2	6	5	1	3	9	8
1	6	9	8	3	2	7	4	5

118

2	7	5	8	1	6	9	4	3
6	9	3	7	5	4	1	8	2
1	8	4	3	2	9	7	6	5
8	5	2	6	4	1	3	7	9
3	4	1	2	9	7	6	5	8
7	6	9	5	8	3	2	1	4
4	1	6	9	3	8	5	2	7
5	3	8	1	7	2	4	9	6
9	2	7	4	6	5	8	3	1

119

5	2	9	3	8	4	7	6	1
1	8	3	2	7	6	4	5	9
6	7	4	1	5	9	3	2	8
9	4	6	7	1	2	5	8	3
8	1	2	5	4	3	6	9	7
7	3	5	6	9	8	2	1	4
4	5	1	9	2	7	8	3	6
2	6	7	8	3	1	9	4	5
3	9	8	4	6	5	1	7	2

120

3	6	2	1	5	4	9	8	7
7	8	5	6	2	9	4	3	1
4	1	9	7	3	8	6	2	5
2	4	6	3	9	7	5	1	8
9	3	1	2	8	5	7	4	6
5	7	8	4	6	1	3	9	2
6	2	4	8	7	3	1	5	9
8	5	3	9	1	6	2	7	4
1	9	7	5	4	2	8	6	3

1

3	2	6	5	7	9	8	1	4
4	9	8	6	1	3	7	5	2
5	1	7	8	2	4	3	6	9
1	5	4	2	8	7	6	9	3
6	8	3	4	9	5	1	2	7
9	7	2	3	6	1	5	4	8
7	6	5	9	4	8	2	3	1
8	3	9	1	5	2	4	7	6
2	4	1	7	3	6	9	8	5

2

9	3	2	6	5	1	4	8	7
4	7	1	2	8	9	6	3	5
5	8	6	4	3	7	1	9	2
7	5	3	8	1	2	9	4	6
2	1	4	3	9	6	7	5	8
8	6	9	7	4	5	3	2	1
1	9	8	5	6	3	2	7	4
3	4	7	1	2	8	5	6	9
6	2	5	9	7	4	8	1	3

3

9	5	7	8	6	3	2	4	1
4	6	2	5	9	1	7	8	3
1	3	8	2	4	7	6	5	9
5	2	1	3	8	6	9	7	4
7	8	4	9	5	2	3	1	6
6	9	3	7	1	4	8	2	5
3	4	9	1	7	8	5	6	2
2	7	6	4	3	5	1	9	8
8	1	5	6	2	9	4	3	7

4

2	1	3	5	6	4	9	8	7
4	7	9	1	3	8	2	5	6
6	8	5	2	7	9	4	1	3
3	6	8	7	9	1	5	4	2
9	4	2	3	8	5	7	6	1
1	5	7	6	4	2	8	3	9
5	3	4	9	1	7	6	2	8
7	2	6	8	5	3	1	9	4
8	9	1	4	2	6	3	7	5

5

8	6	1	5	3	4	9	2	7
2	9	7	6	1	8	4	3	5
5	3	4	2	7	9	6	1	8
4	7	5	1	6	2	3	8	9
6	1	9	4	8	3	5	7	2
3	8	2	9	5	7	1	6	4
9	5	6	7	2	1	8	4	3
7	4	3	8	9	6	2	5	1
1	2	8	3	4	5	7	9	6

6

4	1	2	7	6	3	8	9	5
3	7	8	9	4	5	6	1	2
5	6	9	1	2	8	4	7	3
6	9	4	3	1	2	5	8	7
7	3	1	5	8	6	2	4	9
2	8	5	4	9	7	3	6	1
8	2	7	6	3	1	9	5	4
9	5	3	8	7	4	1	2	6
1	4	6	2	5	9	7	3	8

7

7	1	4	3	2	9	5	8	6
2	3	5	7	8	6	1	9	4
6	8	9	1	5	4	7	2	3
1	5	3	9	6	2	8	4	7
9	4	6	8	1	7	3	5	2
8	2	7	5	4	3	6	1	9
3	7	8	2	9	1	4	6	5
5	6	2	4	3	8	9	7	1
4	9	1	6	7	5	2	3	8

8

4	8	3	2	9	1	6	5	7
5	7	2	8	6	4	3	1	9
6	1	9	3	7	5	2	4	8
7	3	8	9	1	6	4	2	5
2	9	6	4	5	3	7	8	1
1	4	5	7	2	8	9	3	6
9	5	4	1	3	7	8	6	2
8	2	1	6	4	9	5	7	3
3	6	7	5	8	2	1	9	4

9

6	3	4	7	5	8	2	1	9
5	8	9	4	1	2	7	3	6
2	7	1	3	6	9	8	4	5
4	9	2	6	3	5	1	7	8
7	1	5	8	9	4	3	6	2
3	6	8	1	2	7	5	9	4
8	2	6	9	7	3	4	5	1
9	5	7	2	4	1	6	8	3
1	4	3	5	8	6	9	2	7

10

7	8	9	6	5	2	1	3	4
1	3	2	9	4	7	8	6	5
5	4	6	3	8	1	2	7	9
3	6	1	4	7	8	5	9	2
9	2	8	1	6	5	7	4	3
4	5	7	2	9	3	6	8	1
6	9	5	8	1	4	3	2	7
8	1	3	7	2	9	4	5	6
2	7	4	5	3	6	9	1	8

11

5	3	4	7	8	9	6	1	2
2	8	7	1	6	5	4	9	3
1	9	6	4	3	2	7	5	8
8	5	9	6	1	3	2	7	4
7	2	1	8	5	4	3	6	9
6	4	3	2	9	7	1	8	5
9	6	5	3	4	1	8	2	7
3	1	2	9	7	8	5	4	6
4	7	8	5	2	6	9	3	1

12

4	9	3	1	6	7	2	5	8
5	6	7	8	2	9	4	1	3
1	8	2	5	3	4	9	6	7
3	4	8	9	1	6	7	2	5
7	2	6	4	8	5	1	3	9
9	5	1	3	7	2	6	8	4
6	7	5	2	9	8	3	4	1
8	3	9	6	4	1	5	7	2
2	1	4	7	5	3	8	9	6

13

8	7	6	5	1	9	2	4	3
3	2	9	6	7	4	5	8	1
5	4	1	8	2	3	7	9	6
4	8	7	3	6	2	9	1	5
6	1	5	7	9	8	4	3	2
2	9	3	1	4	5	8	6	7
1	3	4	9	5	7	6	2	8
7	6	2	4	8	1	3	5	9
9	5	8	2	3	6	1	7	4

14

8	1	3	4	7	5	9	6	2
5	2	9	8	6	3	4	7	1
7	6	4	1	2	9	8	5	3
2	7	1	3	5	4	6	8	9
6	3	8	9	1	7	2	4	5
9	4	5	6	8	2	1	3	7
1	8	2	5	3	6	7	9	4
3	9	7	2	4	8	5	1	6
4	5	6	7	9	1	3	2	8

15

4	7	8	9	6	3	1	5	2
5	6	1	8	2	4	9	3	7
3	2	9	1	7	5	6	8	4
8	9	4	6	3	7	2	1	5
7	3	2	4	5	1	8	9	6
1	5	6	2	8	9	7	4	3
9	1	5	7	4	6	3	2	8
2	4	7	3	1	8	5	6	9
6	8	3	5	9	2	4	7	1

16

6	8	7	4	2	5	1	3	9
3	9	1	8	7	6	2	5	4
5	2	4	3	9	1	7	8	6
8	4	9	2	5	7	3	6	1
7	3	5	1	6	8	9	4	2
2	1	6	9	4	3	8	7	5
1	6	8	5	3	2	4	9	7
9	7	3	6	1	4	5	2	8
4	5	2	7	8	9	6	1	3

17

5	4	6	9	1	8	2	7	3
9	2	1	3	4	7	6	8	5
3	7	8	5	2	6	1	4	9
4	3	9	2	7	5	8	1	6
1	8	7	6	9	4	5	3	2
6	5	2	1	8	3	4	9	7
2	1	4	7	5	9	3	6	8
8	9	3	4	6	2	7	5	1
7	6	5	8	3	1	9	2	4

18

7	3	1	9	6	4	2	8	5
8	6	4	2	3	5	7	1	9
9	5	2	8	7	1	3	6	4
2	1	5	4	9	6	8	3	7
6	4	7	3	5	8	9	2	1
3	8	9	1	2	7	4	5	6
4	7	6	5	8	3	1	9	2
1	2	8	6	4	9	5	7	3
5	9	3	7	1	2	6	4	8

19

8	9	5	2	6	3	4	1	7
6	2	1	9	7	4	5	3	8
7	4	3	1	5	8	6	9	2
2	3	9	7	4	6	1	8	5
4	8	7	5	1	9	2	6	3
5	1	6	8	3	2	7	4	9
9	7	8	4	2	1	3	5	6
3	5	4	6	9	7	8	2	1
1	6	2	3	8	5	9	7	4

20

4	3	6	1	5	8	7	2	9
1	5	9	4	7	2	3	8	6
7	8	2	3	9	6	1	5	4
8	2	1	7	6	5	4	9	3
9	7	4	8	2	3	5	6	1
3	6	5	9	4	1	8	7	2
2	4	3	6	8	7	9	1	5
5	1	8	2	3	9	6	4	7
6	9	7	5	1	4	2	3	8

21

7	8	9	6	3	2	4	1	5
5	6	3	1	4	7	8	9	2
2	1	4	9	5	8	7	6	3
1	5	2	7	6	9	3	4	8
4	9	8	3	2	1	5	7	6
6	3	7	5	8	4	1	2	9
9	4	5	2	1	3	6	8	7
3	7	1	8	9	6	2	5	4
8	2	6	4	7	5	9	3	1

22

8	6	2	1	5	3	7	4	9
5	9	4	2	7	6	8	1	3
1	3	7	8	9	4	5	2	6
7	4	6	3	8	5	2	9	1
3	1	5	9	2	7	6	8	4
9	2	8	4	6	1	3	5	7
2	7	9	6	4	8	1	3	5
4	5	3	7	1	2	9	6	8
6	8	1	5	3	9	4	7	2

23

7	4	1	5	2	9	6	3	8
8	9	6	3	1	7	2	4	5
2	5	3	8	4	6	7	9	1
1	6	8	2	7	3	9	5	4
5	3	7	6	9	4	1	8	2
4	2	9	1	5	8	3	7	6
9	1	2	4	3	5	8	6	7
3	8	5	7	6	2	4	1	9
6	7	4	9	8	1	5	2	3

24

4	1	5	9	7	6	2	8	3
2	7	6	1	3	8	4	9	5
8	9	3	5	2	4	1	6	7
6	5	4	8	9	2	7	3	1
1	8	7	6	4	3	5	2	9
9	3	2	7	5	1	6	4	8
7	2	1	4	8	9	3	5	6
5	4	8	3	6	7	9	1	2
3	6	9	2	1	5	8	7	4

25

1	9	7	6	8	4	5	3	2
6	8	5	2	3	7	1	4	9
2	4	3	9	1	5	7	6	8
9	1	6	3	4	2	8	5	7
8	3	2	7	5	9	6	1	4
5	7	4	8	6	1	2	9	3
7	2	1	5	9	3	4	8	6
4	6	9	1	2	8	3	7	5
3	5	8	4	7	6	9	2	1

26

2	8	5	3	1	6	7	9	4
4	3	6	8	9	7	2	5	1
7	9	1	4	5	2	6	3	8
1	6	3	9	7	8	4	2	5
9	4	2	6	3	5	8	1	7
8	5	7	2	4	1	9	6	3
5	7	8	1	6	9	3	4	2
3	1	9	7	2	4	5	8	6
6	2	4	5	8	3	1	7	9

27

3	7	9	8	6	4	5	1	2
4	6	1	3	2	5	9	8	7
2	8	5	7	1	9	3	6	4
1	3	8	9	7	2	6	4	5
7	9	4	5	3	6	8	2	1
5	2	6	4	8	1	7	3	9
9	4	2	6	5	8	1	7	3
8	1	7	2	9	3	4	5	6
6	5	3	1	4	7	2	9	8

28

4	3	7	2	8	9	1	6	5
1	6	8	4	3	5	7	9	2
9	2	5	7	6	1	3	8	4
2	8	6	1	7	4	5	3	9
7	9	3	6	5	8	4	2	1
5	1	4	9	2	3	6	7	8
8	5	2	3	1	6	9	4	7
3	7	9	5	4	2	8	1	6
6	4	1	8	9	7	2	5	3

29

7	2	6	1	5	3	4	9	8
4	9	1	8	6	7	5	2	3
3	5	8	4	2	9	6	7	1
2	7	3	9	4	5	8	1	6
8	4	5	6	7	1	2	3	9
6	1	9	2	3	8	7	5	4
5	6	4	3	1	2	9	8	7
9	3	2	7	8	4	1	6	5
1	8	7	5	9	6	3	4	2

30

4	3	8	1	6	2	7	9	5
7	5	6	3	9	8	4	2	1
1	2	9	4	7	5	6	3	8
6	8	4	2	1	9	3	5	7
9	1	3	5	8	7	2	6	4
2	7	5	6	4	3	1	8	9
3	9	1	8	2	4	5	7	6
5	6	7	9	3	1	8	4	2
8	4	2	7	5	6	9	1	3

31

1	5	3	7	9	6	4	2	8
8	4	6	3	5	2	1	9	7
2	7	9	8	4	1	6	5	3
3	9	8	6	1	7	5	4	2
5	2	4	9	3	8	7	6	1
6	1	7	5	2	4	8	3	9
9	6	5	1	8	3	2	7	4
4	3	1	2	7	5	9	8	6
7	8	2	4	6	9	3	1	5

32

8	5	1	4	7	2	6	9	3
9	6	2	8	3	5	1	4	7
4	7	3	1	6	9	8	2	5
6	8	9	5	2	4	7	3	1
1	4	5	3	9	7	2	6	8
2	3	7	6	8	1	4	5	9
7	2	6	9	5	8	3	1	4
5	1	8	2	4	3	9	7	6
3	9	4	7	1	6	5	8	2

33

4	9	3	5	7	1	8	6	2
1	6	8	2	3	4	9	5	7
2	5	7	6	8	9	4	1	3
5	3	2	8	9	7	6	4	1
6	7	4	3	1	5	2	9	8
9	8	1	4	2	6	3	7	5
3	4	9	1	5	8	7	2	6
8	1	6	7	4	2	5	3	9
7	2	5	9	6	3	1	8	4

34

6	3	4	2	5	8	9	1	7
5	1	7	6	4	9	2	3	8
9	8	2	3	7	1	6	5	4
1	9	5	7	8	4	3	6	2
4	2	3	9	1	6	7	8	5
7	6	8	5	2	3	4	9	1
8	4	6	1	3	7	5	2	9
2	7	9	8	6	5	1	4	3
3	5	1	4	9	2	8	7	6

35

3	6	1	5	9	7	4	2	8
7	2	9	4	6	8	1	5	3
4	8	5	2	1	3	7	6	9
6	5	4	8	7	2	3	9	1
9	1	3	6	5	4	8	7	2
2	7	8	1	3	9	6	4	5
8	3	6	7	2	5	9	1	4
1	4	2	9	8	6	5	3	7
5	9	7	3	4	1	2	8	6

36

8	5	6	4	3	7	2	9	1
1	3	7	5	2	9	8	4	6
2	4	9	1	6	8	3	5	7
7	6	4	8	9	3	5	1	2
9	1	5	6	4	2	7	3	8
3	8	2	7	5	1	4	6	9
4	2	1	3	8	6	9	7	5
5	7	8	9	1	4	6	2	3
6	9	3	2	7	5	1	8	4

37

4	1	7	9	6	8	3	5	2
2	3	5	4	1	7	8	6	9
6	8	9	3	2	5	4	7	1
5	7	3	8	4	1	9	2	6
8	9	4	2	5	6	7	1	3
1	6	2	7	3	9	5	4	8
9	5	6	1	7	3	2	8	4
7	2	8	6	9	4	1	3	5
3	4	1	5	8	2	6	9	7

38

5	1	8	7	6	2	9	4	3
4	3	9	8	5	1	2	6	7
6	7	2	3	9	4	8	5	1
9	4	7	5	8	6	3	1	2
8	2	3	1	4	9	6	7	5
1	6	5	2	7	3	4	9	8
7	5	6	4	2	8	1	3	9
2	9	1	6	3	5	7	8	4
3	8	4	9	1	7	5	2	6

39

6	9	4	2	7	3	1	8	5
7	5	3	6	1	8	2	9	4
1	8	2	5	9	4	3	7	6
3	2	9	4	5	7	6	1	8
4	1	5	8	6	2	7	3	9
8	6	7	1	3	9	5	4	2
2	3	1	9	8	6	4	5	7
9	7	6	3	4	5	8	2	1
5	4	8	7	2	1	9	6	3

40

7	9	5	2	4	8	1	3	6
8	2	1	3	9	6	5	7	4
3	4	6	5	7	1	9	2	8
9	5	7	1	2	4	8	6	3
1	8	2	6	3	7	4	9	5
6	3	4	8	5	9	7	1	2
2	7	9	4	6	5	3	8	1
5	6	8	7	1	3	2	4	9
4	1	3	9	8	2	6	5	7

41

3	4	6	9	1	8	7	2	5
7	5	2	6	4	3	9	8	1
9	1	8	2	7	5	4	6	3
5	6	1	3	9	4	8	7	2
8	2	3	7	5	6	1	9	4
4	7	9	8	2	1	5	3	6
1	8	7	4	3	2	6	5	9
2	9	5	1	6	7	3	4	8
6	3	4	5	8	9	2	1	7

42

5	4	6	9	1	3	7	2	8
7	2	3	8	5	6	1	4	9
1	8	9	7	2	4	5	3	6
9	6	8	1	7	2	3	5	4
2	1	4	6	3	5	8	9	7
3	7	5	4	8	9	6	1	2
4	9	7	3	6	1	2	8	5
8	5	1	2	4	7	9	6	3
6	3	2	5	9	8	4	7	1

43

9	7	8	4	1	6	3	5	2
3	6	1	9	5	2	8	7	4
4	5	2	3	7	8	1	9	6
6	2	7	8	9	3	4	1	5
5	4	3	7	2	1	9	6	8
8	1	9	6	4	5	2	3	7
2	3	6	5	8	9	7	4	1
1	9	4	2	6	7	5	8	3
7	8	5	1	3	4	6	2	9

44

3	6	4	2	9	7	5	8	1
8	9	5	3	6	1	4	7	2
2	1	7	8	5	4	3	6	9
7	2	8	5	3	6	1	9	4
5	3	6	4	1	9	8	2	7
9	4	1	7	2	8	6	5	3
6	5	9	1	4	2	7	3	8
1	7	3	9	8	5	2	4	6
4	8	2	6	7	3	9	1	5

45

2	9	7	1	5	4	8	6	3
6	5	1	8	3	2	9	4	7
4	8	3	7	9	6	2	1	5
5	2	6	4	8	9	7	3	1
7	1	9	6	2	3	4	5	8
3	4	8	5	1	7	6	2	9
8	6	4	3	7	1	5	9	2
9	3	5	2	4	8	1	7	6
1	7	2	9	6	5	3	8	4

46

4	6	3	9	8	7	1	5	2
8	7	1	4	5	2	6	9	3
9	2	5	6	3	1	4	7	8
2	3	9	5	7	6	8	4	1
5	1	6	8	4	9	3	2	7
7	8	4	1	2	3	5	6	9
6	9	2	3	1	5	7	8	4
3	4	7	2	6	8	9	1	5
1	5	8	7	9	4	2	3	6

47

3	4	2	7	1	6	9	8	5
1	7	9	5	2	8	6	3	4
5	8	6	4	9	3	1	7	2
6	2	3	9	7	5	4	1	8
8	9	4	6	3	1	5	2	7
7	1	5	2	8	4	3	9	6
9	5	1	8	6	2	7	4	3
4	3	8	1	5	7	2	6	9
2	6	7	3	4	9	8	5	1

48

6	1	9	7	4	2	3	8	5
7	8	2	1	5	3	9	6	4
4	3	5	6	9	8	1	2	7
1	7	4	9	8	6	2	5	3
8	2	6	3	7	5	4	9	1
9	5	3	4	2	1	6	7	8
3	9	8	5	6	4	7	1	2
2	6	1	8	3	7	5	4	9
5	4	7	2	1	9	8	3	6

49

8	6	4	5	1	2	7	3	9
1	3	2	9	6	7	4	8	5
7	9	5	8	3	4	6	1	2
5	2	6	1	9	3	8	7	4
4	8	9	7	2	6	3	5	1
3	1	7	4	5	8	9	2	6
2	5	3	6	8	9	1	4	7
9	4	8	2	7	1	5	6	3
6	7	1	3	4	5	2	9	8

50

2	8	9	7	1	3	5	4	6
5	3	6	9	4	2	1	8	7
7	1	4	8	5	6	2	3	9
6	2	5	1	9	8	4	7	3
1	4	7	3	6	5	8	9	2
3	9	8	2	7	4	6	1	5
9	6	2	4	8	7	3	5	1
4	7	3	5	2	1	9	6	8
8	5	1	6	3	9	7	2	4

51

4	6	7	1	8	3	5	9	2
2	5	3	4	7	9	6	1	8
1	9	8	6	2	5	3	7	4
8	7	1	5	4	6	9	2	3
9	2	6	7	3	1	8	4	5
3	4	5	8	9	2	7	6	1
7	3	4	9	1	8	2	5	6
5	1	2	3	6	7	4	8	9
6	8	9	2	5	4	1	3	7

52

6	5	7	8	1	2	3	4	9
8	4	3	5	9	7	1	6	2
9	1	2	4	6	3	8	7	5
5	7	6	9	3	8	4	2	1
4	9	1	7	2	6	5	8	3
3	2	8	1	5	4	6	9	7
1	8	4	2	7	5	9	3	6
2	6	9	3	8	1	7	5	4
7	3	5	6	4	9	2	1	8

53

3	4	9	6	2	8	1	5	7
8	1	7	5	9	3	4	6	2
2	6	5	4	1	7	3	9	8
9	2	4	1	3	6	7	8	5
7	3	6	8	5	9	2	1	4
5	8	1	7	4	2	9	3	6
6	9	8	3	7	4	5	2	1
4	5	3	2	8	1	6	7	9
1	7	2	9	6	5	8	4	3

54

2	1	6	8	9	3	5	4	7
4	8	7	2	6	5	9	3	1
3	9	5	1	4	7	2	6	8
7	6	1	5	8	4	3	9	2
9	4	2	6	3	1	7	8	5
8	5	3	7	2	9	4	1	6
1	7	9	3	5	8	6	2	4
5	2	4	9	1	6	8	7	3
6	3	8	4	7	2	1	5	9

55

6	3	5	8	2	9	4	7	1
9	1	4	5	7	6	3	8	2
7	2	8	4	3	1	9	6	5
1	6	7	3	9	5	2	4	8
5	8	9	7	4	2	1	3	6
2	4	3	6	1	8	5	9	7
8	9	6	2	5	3	7	1	4
4	5	1	9	6	7	8	2	3
3	7	2	1	8	4	6	5	9

56

2	7	5	9	4	1	8	6	3
4	3	8	6	5	2	7	1	9
1	6	9	7	8	3	2	4	5
3	2	4	1	6	5	9	7	8
5	8	6	4	9	7	3	2	1
9	1	7	3	2	8	6	5	4
8	4	1	2	3	6	5	9	7
6	9	3	5	7	4	1	8	2
7	5	2	8	1	9	4	3	6

57

2	1	6	7	3	5	8	4	9
9	8	7	4	2	1	5	6	3
4	3	5	6	8	9	2	1	7
7	5	1	9	4	3	6	8	2
8	4	9	1	6	2	3	7	5
6	2	3	8	5	7	4	9	1
5	7	4	2	9	6	1	3	8
1	6	2	3	7	8	9	5	4
3	9	8	5	1	4	7	2	6

58

9	4	2	1	5	6	8	7	3
8	3	5	7	4	9	1	2	6
7	6	1	8	2	3	4	9	5
5	2	7	4	9	8	3	6	1
4	9	3	6	7	1	5	8	2
1	8	6	5	3	2	9	4	7
2	7	9	3	8	5	6	1	4
6	5	8	2	1	4	7	3	9
3	1	4	9	6	7	2	5	8

59

5	4	6	8	2	3	7	9	1
7	8	1	5	9	6	3	4	2
3	2	9	4	7	1	5	8	6
8	1	2	3	5	7	9	6	4
4	6	3	9	8	2	1	7	5
9	5	7	6	1	4	2	3	8
2	3	8	1	6	9	4	5	7
6	7	4	2	3	5	8	1	9
1	9	5	7	4	8	6	2	3

60

3	7	5	6	4	9	2	8	1
6	8	1	7	3	2	5	9	4
9	4	2	5	1	8	3	6	7
7	2	9	4	5	1	8	3	6
5	6	3	9	8	7	4	1	2
8	1	4	2	6	3	9	7	5
1	9	7	8	2	4	6	5	3
2	3	6	1	9	5	7	4	8
4	5	8	3	7	6	1	2	9

61

5	7	1	4	6	3	2	9	8
4	9	6	5	8	2	7	1	3
8	3	2	1	9	7	6	4	5
6	1	9	8	7	5	3	2	4
2	4	5	9	3	1	8	6	7
7	8	3	2	4	6	9	5	1
9	2	4	7	1	8	5	3	6
1	6	7	3	5	9	4	8	2
3	5	8	6	2	4	1	7	9

62

5	1	3	8	2	6	4	9	7
2	4	9	7	1	5	6	3	8
6	7	8	4	9	3	2	1	5
3	9	6	1	8	4	7	5	2
7	2	4	6	5	9	3	8	1
8	5	1	2	3	7	9	6	4
1	6	5	9	4	2	8	7	3
9	8	2	3	7	1	5	4	6
4	3	7	5	6	8	1	2	9

63

6	1	4	8	3	5	7	2	9
2	7	5	9	4	1	6	3	8
3	8	9	6	2	7	5	1	4
9	4	3	2	1	6	8	5	7
8	5	7	3	9	4	1	6	2
1	6	2	5	7	8	9	4	3
5	3	1	4	8	9	2	7	6
7	2	8	1	6	3	4	9	5
4	9	6	7	5	2	3	8	1

64

6	4	1	9	7	3	5	2	8
3	9	8	2	5	6	4	1	7
2	7	5	1	4	8	3	6	9
7	8	4	3	9	2	1	5	6
9	6	2	7	1	5	8	3	4
1	5	3	8	6	4	9	7	2
5	1	9	6	8	7	2	4	3
4	3	6	5	2	9	7	8	1
8	2	7	4	3	1	6	9	5

65

6	9	4	7	2	5	1	3	8
7	2	3	1	8	4	9	6	5
8	1	5	9	6	3	2	4	7
9	3	8	5	4	7	6	2	1
5	4	2	6	9	1	7	8	3
1	7	6	2	3	8	4	5	9
4	8	9	3	7	6	5	1	2
3	5	7	4	1	2	8	9	6
2	6	1	8	5	9	3	7	4

66

4	3	8	1	2	6	9	5	7
7	1	2	4	9	5	3	8	6
9	6	5	7	8	3	2	1	4
2	9	1	6	5	8	7	4	3
5	4	3	2	7	9	1	6	8
8	7	6	3	4	1	5	9	2
1	8	7	5	3	4	6	2	9
6	2	9	8	1	7	4	3	5
3	5	4	9	6	2	8	7	1

67

2	9	6	8	7	5	3	1	4
8	5	3	4	9	1	2	6	7
7	1	4	6	2	3	9	8	5
3	7	1	2	5	9	8	4	6
6	8	9	7	3	4	1	5	2
4	2	5	1	8	6	7	3	9
5	3	7	9	6	8	4	2	1
9	4	8	5	1	2	6	7	3
1	6	2	3	4	7	5	9	8

68

9	4	5	8	7	6	3	1	2
8	7	2	3	4	1	5	9	6
3	1	6	2	9	5	8	7	4
2	9	1	7	3	8	6	4	5
6	5	8	4	1	9	7	2	3
4	3	7	5	6	2	1	8	9
7	6	3	9	8	4	2	5	1
1	2	9	6	5	7	4	3	8
5	8	4	1	2	3	9	6	7

69

3	6	8	7	9	1	2	4	5
4	1	5	6	8	2	9	7	3
7	2	9	5	4	3	6	1	8
5	3	1	9	6	4	7	8	2
9	7	6	2	5	8	1	3	4
8	4	2	1	3	7	5	6	9
6	5	3	4	7	9	8	2	1
2	9	4	8	1	6	3	5	7
1	8	7	3	2	5	4	9	6

70

3	6	9	1	2	8	5	4	7
5	8	2	9	4	7	6	3	1
4	7	1	3	6	5	2	8	9
7	2	8	5	9	4	3	1	6
6	4	5	7	1	3	8	9	2
1	9	3	2	8	6	4	7	5
9	5	6	4	3	1	7	2	8
2	3	7	8	5	9	1	6	4
8	1	4	6	7	2	9	5	3

71

4	2	9	7	1	5	6	3	8
7	6	3	8	9	2	4	1	5
8	1	5	3	6	4	2	7	9
6	8	1	4	5	3	7	9	2
9	7	2	1	8	6	5	4	3
3	5	4	9	2	7	8	6	1
2	9	6	5	7	1	3	8	4
5	3	8	6	4	9	1	2	7
1	4	7	2	3	8	9	5	6

72

5	6	2	8	4	7	1	3	9
7	8	9	2	3	1	6	4	5
4	3	1	6	5	9	7	2	8
1	4	6	9	7	2	8	5	3
9	2	3	5	6	8	4	7	1
8	7	5	4	1	3	9	6	2
6	1	7	3	8	5	2	9	4
3	9	8	7	2	4	5	1	6
2	5	4	1	9	6	3	8	7

73

7	1	2	4	8	5	6	3	9
4	9	6	1	7	3	8	2	5
8	5	3	9	6	2	7	4	1
6	3	1	7	2	8	5	9	4
2	8	9	6	5	4	3	1	7
5	7	4	3	9	1	2	6	8
1	4	8	5	3	6	9	7	2
3	2	7	8	4	9	1	5	6
9	6	5	2	1	7	4	8	3

74

6	9	4	3	2	1	8	7	5
2	5	3	7	8	4	9	1	6
7	1	8	5	9	6	2	3	4
5	8	2	6	1	9	3	4	7
9	3	1	8	4	7	5	6	2
4	6	7	2	3	5	1	8	9
8	4	6	1	5	2	7	9	3
1	7	5	9	6	3	4	2	8
3	2	9	4	7	8	6	5	1

75

5	6	1	4	7	3	8	9	2
9	3	7	8	6	2	5	4	1
8	4	2	9	1	5	3	6	7
4	5	8	6	9	1	7	2	3
3	7	6	2	5	4	1	8	9
1	2	9	7	3	8	4	5	6
7	9	5	1	4	6	2	3	8
6	8	4	3	2	7	9	1	5
2	1	3	5	8	9	6	7	4

76

9	5	4	8	2	3	1	7	6
7	8	1	9	4	6	3	2	5
6	3	2	7	1	5	8	4	9
8	7	9	5	6	1	2	3	4
4	2	6	3	8	9	5	1	7
5	1	3	2	7	4	6	9	8
1	4	5	6	3	7	9	8	2
2	9	7	1	5	8	4	6	3
3	6	8	4	9	2	7	5	1

77

1	4	5	6	2	3	9	7	8
6	9	7	8	1	4	5	2	3
2	3	8	7	9	5	1	6	4
7	6	4	9	8	1	3	5	2
3	8	1	2	5	6	4	9	7
5	2	9	4	3	7	8	1	6
4	5	6	3	7	9	2	8	1
9	7	2	1	4	8	6	3	5
8	1	3	5	6	2	7	4	9

78

2	6	4	8	7	9	3	5	1
5	7	1	3	2	6	4	9	8
8	3	9	4	1	5	6	2	7
6	1	5	9	8	4	2	7	3
7	4	8	2	5	3	9	1	6
9	2	3	1	6	7	5	8	4
3	5	2	7	4	8	1	6	9
4	8	6	5	9	1	7	3	2
1	9	7	6	3	2	8	4	5

79

7	9	4	5	8	6	3	1	2
2	6	8	3	1	4	7	9	5
3	5	1	7	9	2	6	8	4
9	1	7	4	6	3	2	5	8
6	8	2	1	5	9	4	3	7
4	3	5	2	7	8	9	6	1
8	4	6	9	2	1	5	7	3
1	7	3	6	4	5	8	2	9
5	2	9	8	3	7	1	4	6

80

3	2	1	7	8	5	9	4	6
8	4	6	9	1	2	5	7	3
9	5	7	4	6	3	8	2	1
1	8	2	3	4	9	7	6	5
4	6	5	1	2	7	3	8	9
7	9	3	8	5	6	4	1	2
6	1	4	5	3	8	2	9	7
2	3	9	6	7	4	1	5	8
5	7	8	2	9	1	6	3	4

81

2	4	1	3	9	7	8	6	5
8	6	3	1	5	2	4	7	9
7	5	9	8	4	6	3	2	1
9	1	5	4	6	3	2	8	7
3	7	4	2	8	9	1	5	6
6	2	8	5	7	1	9	4	3
4	3	7	9	2	5	6	1	8
1	8	6	7	3	4	5	9	2
5	9	2	6	1	8	7	3	4

82

7	9	8	5	3	1	2	4	6
2	5	6	9	4	7	3	8	1
1	3	4	8	2	6	9	7	5
4	8	7	2	1	3	5	6	9
9	1	2	6	5	4	7	3	8
5	6	3	7	8	9	4	1	2
6	2	1	4	7	5	8	9	3
8	7	9	3	6	2	1	5	4
3	4	5	1	9	8	6	2	7

83

1	2	8	7	6	5	4	9	3
4	9	7	2	8	3	6	5	1
3	6	5	4	1	9	7	2	8
6	8	3	9	2	7	1	4	5
7	5	1	3	4	6	2	8	9
2	4	9	8	5	1	3	7	6
9	7	4	1	3	8	5	6	2
5	1	2	6	9	4	8	3	7
8	3	6	5	7	2	9	1	4

84

1	5	9	3	7	6	8	4	2
6	2	4	8	1	5	9	3	7
8	7	3	2	9	4	5	6	1
3	9	6	7	8	2	4	1	5
5	4	1	9	6	3	7	2	8
7	8	2	5	4	1	3	9	6
4	1	7	6	3	8	2	5	9
2	3	8	1	5	9	6	7	4
9	6	5	4	2	7	1	8	3

85

7	6	9	8	2	1	5	4	3
8	5	3	4	7	9	6	2	1
4	1	2	3	5	6	8	9	7
5	4	7	1	6	8	9	3	2
3	9	6	2	4	7	1	5	8
1	2	8	5	9	3	4	7	6
9	8	1	7	3	4	2	6	5
6	3	5	9	8	2	7	1	4
2	7	4	6	1	5	3	8	9

86

2	9	5	1	8	6	3	7	4
7	8	1	3	4	2	6	9	5
6	4	3	7	9	5	1	8	2
1	3	9	2	5	7	4	6	8
4	7	2	8	6	1	5	3	9
8	5	6	9	3	4	7	2	1
9	6	8	5	1	3	2	4	7
3	1	7	4	2	8	9	5	6
5	2	4	6	7	9	8	1	3

87

7	6	3	9	5	8	4	1	2
5	4	8	2	1	6	7	3	9
1	2	9	7	3	4	8	6	5
6	9	5	8	4	3	2	7	1
3	1	4	6	2	7	9	5	8
2	8	7	5	9	1	3	4	6
9	5	1	3	7	2	6	8	4
4	3	6	1	8	9	5	2	7
8	7	2	4	6	5	1	9	3

88

3	2	6	1	8	9	5	7	4
1	9	8	4	5	7	6	3	2
7	5	4	3	6	2	1	8	9
8	6	5	7	2	3	9	4	1
9	7	1	6	4	8	2	5	3
4	3	2	9	1	5	7	6	8
2	4	3	5	9	6	8	1	7
6	8	7	2	3	1	4	9	5
5	1	9	8	7	4	3	2	6

89

2	6	8	1	3	9	4	5	7
3	1	5	7	4	6	8	2	9
7	9	4	5	2	8	6	1	3
1	3	2	9	7	4	5	8	6
5	8	9	6	1	3	7	4	2
6	4	7	8	5	2	9	3	1
9	7	1	2	8	5	3	6	4
4	5	6	3	9	1	2	7	8
8	2	3	4	6	7	1	9	5

90

8	9	2	3	5	6	7	1	4
5	7	6	1	9	4	2	3	8
1	4	3	2	7	8	5	9	6
9	1	5	4	8	7	3	6	2
7	2	8	6	3	1	4	5	9
3	6	4	9	2	5	1	8	7
6	3	1	8	4	2	9	7	5
2	8	7	5	1	9	6	4	3
4	5	9	7	6	3	8	2	1

91

9	8	5	4	2	1	7	6	3
2	3	7	5	6	9	4	8	1
4	1	6	7	3	8	9	5	2
7	6	2	3	4	5	8	1	9
1	5	3	8	9	6	2	4	7
8	4	9	1	7	2	6	3	5
5	9	1	6	8	7	3	2	4
6	2	4	9	5	3	1	7	8
3	7	8	2	1	4	5	9	6

92

4	3	9	8	6	1	5	2	7
8	2	6	7	9	5	3	4	1
7	1	5	3	4	2	8	6	9
5	9	1	6	7	4	2	8	3
3	7	2	5	1	8	6	9	4
6	8	4	2	3	9	1	7	5
2	5	3	4	8	7	9	1	6
1	6	7	9	2	3	4	5	8
9	4	8	1	5	6	7	3	2

93

6	2	5	3	9	1	7	4	8
8	4	3	2	7	6	9	5	1
9	1	7	4	5	8	2	3	6
4	7	6	9	8	5	3	1	2
1	9	2	7	3	4	6	8	5
3	5	8	1	6	2	4	9	7
2	3	9	5	1	7	8	6	4
7	6	1	8	4	3	5	2	9
5	8	4	6	2	9	1	7	3

94

5	6	8	3	9	2	7	4	1
9	1	4	6	7	8	2	3	5
7	3	2	5	4	1	6	8	9
2	4	1	8	3	7	9	5	6
3	7	9	4	6	5	8	1	2
6	8	5	1	2	9	3	7	4
1	5	7	2	8	6	4	9	3
8	2	3	9	1	4	5	6	7
4	9	6	7	5	3	1	2	8

95

7	6	1	9	4	3	2	8	5
3	9	4	5	8	2	7	1	6
5	8	2	1	6	7	9	3	4
1	5	9	3	2	4	8	6	7
2	7	8	6	9	5	1	4	3
4	3	6	7	1	8	5	9	2
9	2	3	8	7	6	4	5	1
8	4	5	2	3	1	6	7	9
6	1	7	4	5	9	3	2	8

96

5	4	6	3	8	1	9	2	7
1	3	9	2	7	6	4	8	5
7	2	8	9	4	5	1	3	6
2	9	1	8	6	4	7	5	3
3	6	5	7	1	2	8	9	4
8	7	4	5	3	9	6	1	2
4	5	7	1	2	8	3	6	9
6	1	2	4	9	3	5	7	8
9	8	3	6	5	7	2	4	1

97

3	4	8	7	5	2	1	9	6
2	1	5	3	6	9	7	4	8
9	6	7	8	1	4	5	3	2
1	8	3	2	9	7	6	5	4
4	9	2	6	3	5	8	1	7
7	5	6	1	4	8	9	2	3
8	7	1	5	2	3	4	6	9
6	3	4	9	8	1	2	7	5
5	2	9	4	7	6	3	8	1

98

4	9	2	1	8	3	7	6	5
7	3	1	5	4	6	9	8	2
5	8	6	7	9	2	1	3	4
8	1	5	2	3	9	6	4	7
9	7	4	6	5	8	2	1	3
2	6	3	4	7	1	5	9	8
1	5	8	9	2	4	3	7	6
6	4	7	3	1	5	8	2	9
3	2	9	8	6	7	4	5	1

99

6	4	9	1	2	8	7	5	3
2	1	3	5	4	7	9	6	8
8	7	5	6	3	9	4	2	1
1	5	2	7	6	4	8	3	9
3	6	4	9	8	1	2	7	5
7	9	8	3	5	2	1	4	6
9	3	1	4	7	6	5	8	2
4	2	6	8	9	5	3	1	7
5	8	7	2	1	3	6	9	4

100

4	3	8	2	5	6	7	9	1
6	7	5	1	4	9	3	2	8
2	9	1	7	8	3	6	4	5
7	8	3	5	1	2	9	6	4
1	4	9	8	6	7	5	3	2
5	6	2	3	9	4	8	1	7
3	5	6	4	7	1	2	8	9
9	1	7	6	2	8	4	5	3
8	2	4	9	3	5	1	7	6

101

2	3	9	1	8	4	5	7	6
7	8	4	5	6	2	9	1	3
5	6	1	7	3	9	8	2	4
3	5	8	6	9	1	2	4	7
6	1	2	4	7	8	3	5	9
4	9	7	3	2	5	6	8	1
9	7	5	8	4	3	1	6	2
1	2	6	9	5	7	4	3	8
8	4	3	2	1	6	7	9	5

102

5	2	1	9	7	4	6	8	3
8	9	6	5	3	2	7	4	1
7	3	4	8	1	6	2	5	9
4	6	9	7	8	1	3	2	5
3	1	8	2	6	5	4	9	7
2	5	7	3	4	9	8	1	6
9	4	2	6	5	3	1	7	8
1	7	3	4	9	8	5	6	2
6	8	5	1	2	7	9	3	4

103

5	1	6	4	9	8	7	2	3
2	8	4	7	3	5	1	9	6
7	9	3	1	6	2	5	4	8
9	6	1	8	4	3	2	7	5
3	7	8	5	2	9	4	6	1
4	2	5	6	1	7	8	3	9
6	5	2	9	8	4	3	1	7
8	4	9	3	7	1	6	5	2
1	3	7	2	5	6	9	8	4

104

4	3	5	8	7	6	1	9	2
2	6	1	3	9	4	8	7	5
9	8	7	2	1	5	6	4	3
3	7	2	9	6	8	5	1	4
6	5	9	1	4	7	2	3	8
1	4	8	5	2	3	7	6	9
7	9	4	6	8	2	3	5	1
5	2	6	4	3	1	9	8	7
8	1	3	7	5	9	4	2	6

105

2	5	8	6	4	3	9	1	7
3	7	4	9	2	1	8	5	6
9	1	6	5	7	8	3	4	2
6	2	5	8	9	4	7	3	1
8	4	3	7	1	5	2	6	9
7	9	1	2	3	6	5	8	4
1	3	2	4	5	9	6	7	8
4	8	7	3	6	2	1	9	5
5	6	9	1	8	7	4	2	3

106

7	5	4	3	1	2	9	6	8
2	6	1	9	4	8	5	3	7
3	9	8	6	5	7	1	4	2
8	4	3	1	2	5	6	7	9
1	7	6	4	3	9	2	8	5
5	2	9	7	8	6	4	1	3
4	3	2	5	7	1	8	9	6
6	1	5	8	9	3	7	2	4
9	8	7	2	6	4	3	5	1

107

4	1	6	3	9	5	8	2	7
8	5	3	1	2	7	9	4	6
2	9	7	6	4	8	1	3	5
5	6	9	8	1	4	2	7	3
7	8	2	9	3	6	5	1	4
1	3	4	7	5	2	6	8	9
9	7	1	5	8	3	4	6	2
3	4	5	2	6	1	7	9	8
6	2	8	4	7	9	3	5	1

108

2	5	6	4	1	8	3	9	7
7	1	8	6	9	3	5	4	2
4	3	9	2	7	5	6	1	8
8	2	5	7	4	6	1	3	9
3	9	4	5	8	1	7	2	6
6	7	1	9	3	2	8	5	4
5	4	7	1	6	9	2	8	3
9	8	2	3	5	7	4	6	1
1	6	3	8	2	4	9	7	5

109

5	4	7	2	1	8	6	3	9
6	1	9	4	3	5	8	7	2
2	3	8	6	9	7	5	1	4
9	7	6	5	4	1	2	8	3
4	2	1	3	8	6	7	9	5
8	5	3	9	7	2	1	4	6
1	6	4	7	2	9	3	5	8
7	9	2	8	5	3	4	6	1
3	8	5	1	6	4	9	2	7

110

9	6	3	5	7	2	8	4	1
2	1	5	4	3	8	6	9	7
7	4	8	6	9	1	2	5	3
4	9	6	3	1	7	5	8	2
3	8	7	2	5	4	1	6	9
1	5	2	8	6	9	7	3	4
6	7	1	9	4	5	3	2	8
8	3	9	7	2	6	4	1	5
5	2	4	1	8	3	9	7	6

111

9	6	1	7	8	5	2	4	3
8	2	5	4	9	3	6	7	1
7	3	4	2	6	1	9	5	8
5	8	6	9	3	4	7	1	2
4	1	7	6	2	8	5	3	9
2	9	3	5	1	7	8	6	4
3	7	2	8	4	6	1	9	5
6	4	8	1	5	9	3	2	7
1	5	9	3	7	2	4	8	6

112

2	4	7	1	3	9	6	5	8
6	1	8	5	2	7	3	9	4
3	9	5	8	4	6	1	2	7
7	6	1	4	8	2	5	3	9
4	3	9	6	1	5	7	8	2
5	8	2	7	9	3	4	1	6
1	7	6	9	5	8	2	4	3
9	5	3	2	6	4	8	7	1
8	2	4	3	7	1	9	6	5

113

6	4	8	2	7	1	3	9	5
5	2	7	3	6	9	8	1	4
1	9	3	5	8	4	7	2	6
2	8	5	6	4	7	1	3	9
4	1	6	9	3	8	5	7	2
7	3	9	1	5	2	4	6	8
9	7	2	8	1	5	6	4	3
8	6	4	7	9	3	2	5	1
3	5	1	4	2	6	9	8	7

114

3	2	8	9	5	7	4	1	6
9	5	1	6	3	4	2	7	8
4	6	7	8	2	1	3	9	5
7	3	6	4	8	9	5	2	1
8	9	2	7	1	5	6	3	4
1	4	5	3	6	2	9	8	7
6	1	9	2	4	8	7	5	3
2	8	4	5	7	3	1	6	9
5	7	3	1	9	6	8	4	2

115

7	1	6	4	3	2	8	9	5
8	2	5	1	9	7	4	3	6
3	9	4	5	8	6	7	1	2
2	4	3	8	5	1	6	7	9
1	6	8	3	7	9	2	5	4
5	7	9	6	2	4	1	8	3
4	8	1	9	6	3	5	2	7
9	5	7	2	4	8	3	6	1
6	3	2	7	1	5	9	4	8

116

5	8	4	6	7	3	2	1	9
3	7	1	9	5	2	8	6	4
2	6	9	4	1	8	5	3	7
1	4	8	2	6	7	3	9	5
7	2	5	3	9	4	6	8	1
9	3	6	5	8	1	4	7	2
6	5	3	7	2	9	1	4	8
8	9	2	1	4	6	7	5	3
4	1	7	8	3	5	9	2	6

117

7	3	4	8	9	5	6	1	2
9	2	5	7	6	1	8	4	3
6	1	8	3	2	4	5	7	9
2	8	3	6	7	9	4	5	1
5	6	7	4	1	3	9	2	8
4	9	1	5	8	2	3	6	7
3	7	9	1	5	6	2	8	4
8	5	2	9	4	7	1	3	6
1	4	6	2	3	8	7	9	5

118

6	1	3	7	2	5	8	9	4
5	7	8	9	4	1	2	6	3
4	9	2	3	6	8	5	7	1
8	2	7	6	1	3	9	4	5
3	5	4	2	8	9	6	1	7
9	6	1	5	7	4	3	2	8
7	8	9	1	3	2	4	5	6
1	4	5	8	9	6	7	3	2
2	3	6	4	5	7	1	8	9

119

3	1	7	8	4	5	6	2	9
8	4	5	9	2	6	7	1	3
6	9	2	1	3	7	4	5	8
9	5	6	2	1	8	3	7	4
4	3	1	7	5	9	2	8	6
7	2	8	3	6	4	1	9	5
2	7	9	4	8	3	5	6	1
1	6	4	5	9	2	8	3	7
5	8	3	6	7	1	9	4	2

120

8	3	2	4	6	1	5	7	9
9	4	1	5	7	8	2	6	3
7	5	6	9	2	3	1	8	4
6	7	9	1	5	2	4	3	8
5	1	3	8	4	9	7	2	6
4	2	8	7	3	6	9	5	1
1	9	5	6	8	7	3	4	2
2	8	4	3	9	5	6	1	7
3	6	7	2	1	4	8	9	5

1

5	6	2	3	9	7	4	1	8
8	7	9	1	4	6	2	3	5
4	1	3	8	5	2	7	6	9
9	4	6	7	8	5	1	2	3
1	2	8	6	3	9	5	7	4
7	3	5	4	2	1	8	9	6
2	8	4	9	7	3	6	5	1
3	5	1	2	6	8	9	4	7
6	9	7	5	1	4	3	8	2

2

7	5	2	9	6	8	3	4	1
1	3	6	4	7	5	2	9	8
4	9	8	3	1	2	7	5	6
9	7	4	1	2	6	5	8	3
8	2	1	5	3	7	4	6	9
3	6	5	8	9	4	1	2	7
2	4	9	7	8	3	6	1	5
6	8	3	2	5	1	9	7	4
5	1	7	6	4	9	8	3	2

3

3	1	4	8	7	2	6	9	5
2	8	7	6	5	9	1	3	4
6	9	5	3	1	4	8	7	2
1	2	6	9	3	8	5	4	7
4	5	3	7	6	1	9	2	8
8	7	9	2	4	5	3	6	1
9	3	1	4	8	7	2	5	6
5	4	2	1	9	6	7	8	3
7	6	8	5	2	3	4	1	9

4

9	8	2	5	7	6	1	3	4
5	1	6	2	3	4	7	9	8
3	7	4	1	8	9	2	6	5
4	6	7	3	2	1	8	5	9
8	9	3	7	4	5	6	1	2
2	5	1	6	9	8	4	7	3
1	3	9	4	6	2	5	8	7
7	4	5	8	1	3	9	2	6
6	2	8	9	5	7	3	4	1

5

2	5	3	8	9	6	1	4	7
6	7	1	5	4	3	8	9	2
8	4	9	2	7	1	5	6	3
7	3	6	4	2	8	9	5	1
5	2	8	6	1	9	7	3	4
1	9	4	3	5	7	2	8	6
4	8	2	1	3	5	6	7	9
3	6	7	9	8	2	4	1	5
9	1	5	7	6	4	3	2	8

6

5	6	8	9	1	4	2	3	7
9	2	3	7	6	5	8	1	4
7	4	1	3	8	2	9	6	5
2	7	4	6	3	1	5	8	9
1	8	5	2	9	7	3	4	6
6	3	9	4	5	8	7	2	1
8	5	6	1	7	3	4	9	2
3	1	2	5	4	9	6	7	8
4	9	7	8	2	6	1	5	3

7

8	7	3	4	9	5	1	2	6
5	1	9	2	6	8	3	7	4
4	2	6	7	3	1	9	5	8
6	9	2	8	1	3	5	4	7
1	8	4	6	5	7	2	9	3
3	5	7	9	4	2	8	6	1
7	4	8	1	2	9	6	3	5
9	3	1	5	7	6	4	8	2
2	6	5	3	8	4	7	1	9

8

1	7	6	3	5	8	9	2	4
2	3	9	4	6	7	1	8	5
4	8	5	9	2	1	3	6	7
5	9	7	6	4	3	2	1	8
8	1	3	2	7	5	6	4	9
6	2	4	1	8	9	7	5	3
3	4	1	5	9	2	8	7	6
7	5	2	8	3	6	4	9	1
9	6	8	7	1	4	5	3	2

9

6	1	4	5	9	8	2	3	7
3	9	2	7	4	1	5	6	8
8	5	7	6	2	3	9	4	1
4	6	9	3	8	2	7	1	5
2	8	5	1	6	7	3	9	4
7	3	1	4	5	9	6	8	2
5	7	3	9	1	4	8	2	6
9	4	8	2	7	6	1	5	3
1	2	6	8	3	5	4	7	9

10

4	2	3	9	6	1	8	5	7
5	6	1	2	8	7	4	9	3
8	7	9	5	4	3	1	6	2
3	1	2	7	9	4	5	8	6
9	4	6	3	5	8	2	7	1
7	5	8	6	1	2	9	3	4
6	3	4	8	2	5	7	1	9
2	8	7	1	3	9	6	4	5
1	9	5	4	7	6	3	2	8

11

9	6	8	1	5	4	3	2	7
7	4	1	2	6	3	9	5	8
2	3	5	8	7	9	6	4	1
8	2	6	3	1	7	5	9	4
4	9	7	5	2	6	8	1	3
1	5	3	4	9	8	2	7	6
6	1	9	7	3	5	4	8	2
5	7	4	6	8	2	1	3	9
3	8	2	9	4	1	7	6	5

12

8	5	2	1	6	7	9	4	3
9	6	3	5	8	4	7	2	1
7	4	1	2	9	3	8	6	5
2	3	5	6	7	9	1	8	4
6	8	7	4	1	5	3	9	2
4	1	9	8	3	2	6	5	7
1	2	6	7	4	8	5	3	9
5	9	8	3	2	1	4	7	6
3	7	4	9	5	6	2	1	8

13

1	6	9	5	7	2	4	8	3
3	8	7	9	1	4	2	5	6
2	4	5	6	8	3	9	1	7
4	5	2	7	3	8	6	9	1
7	9	6	1	4	5	3	2	8
8	1	3	2	6	9	7	4	5
6	7	4	8	9	1	5	3	2
9	2	8	3	5	6	1	7	4
5	3	1	4	2	7	8	6	9

14

1	9	3	5	4	8	7	2	6
8	5	6	7	1	2	3	9	4
2	4	7	3	9	6	8	1	5
5	6	9	8	3	7	2	4	1
3	7	2	4	5	1	6	8	9
4	1	8	6	2	9	5	3	7
9	3	4	2	6	5	1	7	8
6	8	1	9	7	3	4	5	2
7	2	5	1	8	4	9	6	3

15

4	1	8	9	2	5	7	6	3
3	9	7	6	8	1	4	5	2
6	2	5	7	3	4	9	8	1
9	7	1	2	5	6	3	4	8
8	6	2	4	9	3	5	1	7
5	3	4	8	1	7	6	2	9
1	4	9	5	7	8	2	3	6
2	5	3	1	6	9	8	7	4
7	8	6	3	4	2	1	9	5

16

5	7	8	1	3	6	4	2	9
9	4	6	8	5	2	1	7	3
3	2	1	7	4	9	5	6	8
4	1	2	9	8	7	3	5	6
6	3	9	4	2	5	7	8	1
7	8	5	3	6	1	9	4	2
8	9	3	6	7	4	2	1	5
2	6	7	5	1	3	8	9	4
1	5	4	2	9	8	6	3	7

17

9	7	3	1	4	6	2	5	8
8	5	6	3	2	7	1	9	4
4	1	2	8	9	5	3	7	6
5	6	7	4	8	1	9	3	2
3	2	4	5	6	9	8	1	7
1	9	8	7	3	2	4	6	5
2	3	1	6	7	4	5	8	9
7	8	9	2	5	3	6	4	1
6	4	5	9	1	8	7	2	3

18

2	8	7	3	4	9	6	5	1
4	5	3	6	1	8	2	7	9
6	9	1	7	2	5	3	4	8
7	2	5	8	9	1	4	6	3
9	3	4	5	6	7	8	1	2
8	1	6	2	3	4	7	9	5
3	6	9	1	7	2	5	8	4
5	4	2	9	8	6	1	3	7
1	7	8	4	5	3	9	2	6

19

5	9	8	6	4	3	7	2	1
4	6	2	1	7	8	3	5	9
1	3	7	9	5	2	8	4	6
9	1	4	7	8	6	5	3	2
7	5	6	3	2	1	9	8	4
2	8	3	4	9	5	6	1	7
8	7	5	2	6	4	1	9	3
6	2	1	8	3	9	4	7	5
3	4	9	5	1	7	2	6	8

20

3	4	8	6	1	7	9	2	5
6	2	5	4	9	3	8	1	7
7	9	1	8	2	5	3	6	4
2	1	7	9	3	6	5	4	8
8	5	3	1	4	2	6	7	9
4	6	9	7	5	8	1	3	2
5	7	2	3	8	1	4	9	6
9	3	6	5	7	4	2	8	1
1	8	4	2	6	9	7	5	3

21

4	7	2	3	6	1	5	8	9
1	8	3	4	9	5	7	2	6
9	5	6	7	8	2	3	4	1
7	6	1	5	4	9	2	3	8
2	3	8	1	7	6	9	5	4
5	9	4	2	3	8	1	6	7
8	2	5	6	1	7	4	9	3
6	4	7	9	5	3	8	1	2
3	1	9	8	2	4	6	7	5

22

9	7	4	3	5	2	8	6	1
2	5	1	6	4	8	7	3	9
3	6	8	7	9	1	5	4	2
4	3	7	8	1	5	2	9	6
5	1	9	2	7	6	4	8	3
8	2	6	4	3	9	1	5	7
7	8	5	9	2	3	6	1	4
1	9	2	5	6	4	3	7	8
6	4	3	1	8	7	9	2	5

23

1	5	4	2	8	3	9	7	6
3	2	9	1	6	7	8	5	4
7	6	8	5	4	9	2	1	3
4	7	2	8	9	5	6	3	1
5	3	6	7	1	2	4	9	8
8	9	1	4	3	6	5	2	7
6	1	7	9	5	8	3	4	2
2	8	5	3	7	4	1	6	9
9	4	3	6	2	1	7	8	5

24

5	8	9	1	7	4	2	3	6
2	4	7	6	5	3	9	8	1
1	6	3	8	9	2	4	5	7
6	1	2	9	4	8	5	7	3
4	3	5	7	2	6	1	9	8
9	7	8	5	3	1	6	2	4
7	5	1	3	6	9	8	4	2
3	2	6	4	8	5	7	1	9
8	9	4	2	1	7	3	6	5

25

2	8	7	1	5	9	3	6	4
5	3	6	7	4	8	1	9	2
1	9	4	6	2	3	8	5	7
7	4	9	5	3	2	6	1	8
3	1	2	9	8	6	4	7	5
6	5	8	4	7	1	2	3	9
8	2	5	3	6	7	9	4	1
4	6	1	2	9	5	7	8	3
9	7	3	8	1	4	5	2	6

26

4	1	8	9	2	5	6	3	7
7	2	6	3	8	1	9	4	5
3	5	9	4	7	6	8	1	2
8	9	2	7	4	3	1	5	6
5	4	3	1	6	8	7	2	9
1	6	7	5	9	2	3	8	4
9	7	1	8	5	4	2	6	3
2	8	5	6	3	9	4	7	1
6	3	4	2	1	7	5	9	8

27

2	3	5	9	4	7	8	6	1
8	4	7	2	1	6	5	3	9
6	1	9	8	3	5	4	7	2
3	7	1	5	9	8	6	2	4
9	5	8	6	2	4	3	1	7
4	6	2	1	7	3	9	8	5
7	9	3	4	8	1	2	5	6
1	2	6	3	5	9	7	4	8
5	8	4	7	6	2	1	9	3

28

1	2	7	8	9	5	4	6	3
4	6	3	2	7	1	8	5	9
8	9	5	6	4	3	2	7	1
5	1	4	3	8	7	6	9	2
6	3	9	5	2	4	1	8	7
7	8	2	9	1	6	5	3	4
9	7	1	4	6	8	3	2	5
2	5	6	1	3	9	7	4	8
3	4	8	7	5	2	9	1	6

29

1	7	3	2	4	5	9	8	6
6	9	8	3	7	1	2	4	5
5	2	4	9	6	8	3	7	1
2	4	9	1	5	6	8	3	7
3	6	1	8	9	7	5	2	4
8	5	7	4	3	2	1	6	9
7	1	6	5	8	3	4	9	2
9	8	2	7	1	4	6	5	3
4	3	5	6	2	9	7	1	8

30

2	8	4	5	6	1	7	3	9
6	1	9	7	2	3	5	8	4
3	5	7	8	4	9	1	2	6
7	9	2	4	1	5	8	6	3
5	6	1	9	3	8	4	7	2
8	4	3	2	7	6	9	5	1
1	7	5	6	9	2	3	4	8
9	2	8	3	5	4	6	1	7
4	3	6	1	8	7	2	9	5

31

4	3	2	8	9	1	7	5	6
9	7	5	4	3	6	1	2	8
6	1	8	5	7	2	4	9	3
7	8	1	2	4	9	6	3	5
3	9	4	7	6	5	2	8	1
2	5	6	1	8	3	9	7	4
1	2	7	6	5	8	3	4	9
8	4	3	9	1	7	5	6	2
5	6	9	3	2	4	8	1	7

32

2	7	3	8	9	6	1	4	5
8	6	1	3	5	4	9	7	2
9	4	5	2	1	7	3	6	8
6	2	4	9	7	5	8	1	3
3	9	7	6	8	1	5	2	4
5	1	8	4	2	3	7	9	6
7	3	6	5	4	9	2	8	1
4	8	9	1	3	2	6	5	7
1	5	2	7	6	8	4	3	9

33

2	8	6	7	9	1	5	3	4
4	7	3	2	5	8	6	1	9
5	9	1	3	6	4	2	7	8
7	1	5	9	4	3	8	2	6
8	4	2	5	1	6	3	9	7
6	3	9	8	2	7	4	5	1
3	5	8	6	7	9	1	4	2
9	6	4	1	3	2	7	8	5
1	2	7	4	8	5	9	6	3

34

1	6	9	5	2	3	4	8	7
5	7	4	6	8	9	2	3	1
8	3	2	1	4	7	6	5	9
7	4	6	8	3	2	1	9	5
3	8	5	9	1	4	7	6	2
2	9	1	7	6	5	8	4	3
6	1	3	2	9	8	5	7	4
9	5	8	4	7	1	3	2	6
4	2	7	3	5	6	9	1	8

35

2	5	3	1	6	8	7	4	9
1	4	7	2	3	9	5	8	6
6	9	8	5	4	7	3	2	1
3	7	9	8	5	2	6	1	4
4	6	1	7	9	3	2	5	8
8	2	5	6	1	4	9	3	7
9	8	2	3	7	1	4	6	5
7	3	6	4	8	5	1	9	2
5	1	4	9	2	6	8	7	3

36

5	6	2	3	4	8	9	7	1
9	8	1	2	6	7	3	4	5
3	7	4	5	9	1	6	2	8
7	5	3	1	8	9	4	6	2
2	9	8	4	3	6	5	1	7
4	1	6	7	2	5	8	3	9
1	3	5	9	7	4	2	8	6
8	2	9	6	1	3	7	5	4
6	4	7	8	5	2	1	9	3

37

6	3	7	8	1	5	9	4	2
9	8	4	3	7	2	5	1	6
5	2	1	6	9	4	7	8	3
8	4	5	2	3	6	1	7	9
1	9	2	7	4	8	6	3	5
3	7	6	1	5	9	4	2	8
2	1	9	5	8	7	3	6	4
7	5	8	4	6	3	2	9	1
4	6	3	9	2	1	8	5	7

38

8	6	1	7	4	2	3	5	9
2	4	9	8	3	5	7	1	6
3	7	5	6	1	9	2	8	4
4	1	6	3	9	8	5	2	7
5	2	8	4	7	1	6	9	3
7	9	3	2	5	6	8	4	1
9	3	2	1	8	7	4	6	5
1	8	4	5	6	3	9	7	2
6	5	7	9	2	4	1	3	8

39

8	3	6	9	1	4	2	5	7
5	1	7	8	2	6	9	4	3
2	9	4	3	7	5	1	8	6
9	5	2	6	3	8	7	1	4
6	8	1	4	9	7	5	3	2
4	7	3	1	5	2	8	6	9
1	4	5	2	6	9	3	7	8
7	6	9	5	8	3	4	2	1
3	2	8	7	4	1	6	9	5

40

5	2	4	6	7	8	9	3	1
1	9	3	5	4	2	7	6	8
6	8	7	3	1	9	4	5	2
4	6	1	9	2	7	5	8	3
9	7	2	8	3	5	1	4	6
8	3	5	4	6	1	2	7	9
2	1	6	7	5	3	8	9	4
3	5	9	2	8	4	6	1	7
7	4	8	1	9	6	3	2	5

41

8	6	2	1	4	3	5	9	7
1	7	4	5	9	6	2	8	3
3	5	9	2	7	8	6	4	1
9	2	3	4	8	7	1	6	5
6	4	1	9	2	5	7	3	8
7	8	5	3	6	1	4	2	9
4	1	8	6	5	9	3	7	2
5	9	6	7	3	2	8	1	4
2	3	7	8	1	4	9	5	6

42

1	9	8	3	7	5	4	6	2
2	6	4	8	9	1	7	5	3
5	3	7	4	6	2	1	9	8
7	4	5	2	8	9	6	3	1
9	1	2	6	3	7	5	8	4
3	8	6	1	5	4	2	7	9
6	5	1	9	4	3	8	2	7
4	7	3	5	2	8	9	1	6
8	2	9	7	1	6	3	4	5

43

7	8	4	1	3	6	9	2	5
1	3	5	2	9	7	4	6	8
9	2	6	5	8	4	7	1	3
2	7	3	8	1	9	5	4	6
6	9	1	4	5	2	3	8	7
4	5	8	6	7	3	1	9	2
8	6	7	3	4	1	2	5	9
5	4	9	7	2	8	6	3	1
3	1	2	9	6	5	8	7	4

44

6	2	3	5	7	9	4	1	8
5	9	4	8	3	1	6	7	2
7	1	8	4	2	6	9	5	3
4	5	9	7	1	8	3	2	6
8	6	7	2	5	3	1	9	4
2	3	1	6	9	4	7	8	5
1	4	6	9	8	2	5	3	7
3	8	5	1	4	7	2	6	9
9	7	2	3	6	5	8	4	1

45

7	6	2	9	4	3	1	5	8
8	4	3	1	5	2	7	9	6
1	9	5	7	8	6	2	3	4
4	2	9	8	6	1	5	7	3
3	5	1	4	9	7	6	8	2
6	8	7	3	2	5	4	1	9
9	1	8	2	7	4	3	6	5
2	7	6	5	3	9	8	4	1
5	3	4	6	1	8	9	2	7

46

7	1	3	4	2	5	6	9	8
4	9	8	3	7	6	2	5	1
6	2	5	1	8	9	4	7	3
1	4	2	5	6	8	7	3	9
9	5	6	7	4	3	1	8	2
8	3	7	9	1	2	5	4	6
3	7	4	2	9	1	8	6	5
5	6	1	8	3	4	9	2	7
2	8	9	6	5	7	3	1	4

47

1	3	9	6	5	7	4	8	2
4	6	5	9	2	8	3	1	7
7	2	8	1	4	3	9	6	5
9	1	3	2	7	5	6	4	8
2	7	4	8	1	6	5	3	9
5	8	6	3	9	4	2	7	1
3	4	1	5	8	9	7	2	6
8	9	7	4	6	2	1	5	3
6	5	2	7	3	1	8	9	4

48

5	8	9	7	4	2	3	6	1
7	2	1	8	6	3	5	9	4
4	3	6	5	9	1	2	8	7
1	9	2	6	3	5	7	4	8
8	7	5	9	1	4	6	2	3
3	6	4	2	8	7	1	5	9
9	5	3	4	7	6	8	1	2
2	1	8	3	5	9	4	7	6
6	4	7	1	2	8	9	3	5

49

5	9	1	8	4	7	3	6	2
6	7	4	2	3	9	1	5	8
3	8	2	1	6	5	4	7	9
7	3	5	6	9	4	2	8	1
4	2	6	7	8	1	9	3	5
8	1	9	5	2	3	7	4	6
9	5	8	4	7	2	6	1	3
1	4	3	9	5	6	8	2	7
2	6	7	3	1	8	5	9	4

50

9	4	7	6	8	5	1	3	2
8	5	2	4	1	3	6	7	9
1	3	6	2	7	9	5	8	4
3	6	1	8	9	7	2	4	5
5	8	9	3	4	2	7	1	6
7	2	4	5	6	1	3	9	8
6	1	5	9	3	4	8	2	7
2	9	3	7	5	8	4	6	1
4	7	8	1	2	6	9	5	3

51

2	9	8	7	3	6	4	5	1
7	6	5	1	4	9	8	3	2
4	1	3	8	5	2	7	9	6
5	4	6	3	1	7	2	8	9
9	8	7	6	2	5	3	1	4
1	3	2	9	8	4	6	7	5
3	7	4	2	9	1	5	6	8
6	5	9	4	7	8	1	2	3
8	2	1	5	6	3	9	4	7

52

2	6	1	5	9	3	8	7	4
9	4	5	7	6	8	1	3	2
7	3	8	2	4	1	6	5	9
1	5	6	4	7	2	3	9	8
4	9	3	8	5	6	7	2	1
8	7	2	1	3	9	5	4	6
3	1	4	9	8	7	2	6	5
6	2	9	3	1	5	4	8	7
5	8	7	6	2	4	9	1	3

53

3	7	8	4	5	2	1	9	6
6	9	4	8	7	1	3	2	5
1	2	5	6	9	3	4	7	8
8	1	6	7	3	4	2	5	9
9	5	7	2	1	8	6	3	4
4	3	2	5	6	9	8	1	7
7	6	3	1	4	5	9	8	2
5	8	9	3	2	6	7	4	1
2	4	1	9	8	7	5	6	3

54

4	9	1	6	3	2	5	8	7
5	2	6	7	1	8	3	9	4
7	3	8	5	9	4	6	1	2
8	5	4	9	7	3	2	6	1
3	6	7	2	4	1	8	5	9
9	1	2	8	5	6	4	7	3
2	8	3	1	6	7	9	4	5
1	4	5	3	8	9	7	2	6
6	7	9	4	2	5	1	3	8

55

2	8	1	9	7	5	4	3	6
5	9	6	8	4	3	7	2	1
4	3	7	6	2	1	9	5	8
1	2	9	3	8	6	5	4	7
7	5	3	1	9	4	8	6	2
8	6	4	7	5	2	1	9	3
9	1	2	5	3	8	6	7	4
3	7	8	4	6	9	2	1	5
6	4	5	2	1	7	3	8	9

56

5	4	2	1	9	7	6	3	8
3	6	7	5	4	8	9	1	2
9	8	1	3	6	2	7	5	4
8	3	5	6	1	4	2	9	7
4	2	6	9	7	3	1	8	5
7	1	9	8	2	5	3	4	6
2	7	3	4	5	1	8	6	9
6	5	8	2	3	9	4	7	1
1	9	4	7	8	6	5	2	3

57

6	7	4	8	9	3	5	2	1
2	8	5	1	4	7	9	3	6
9	3	1	2	5	6	8	7	4
7	2	8	3	1	5	4	6	9
1	6	9	7	2	4	3	8	5
5	4	3	6	8	9	7	1	2
3	9	7	5	6	2	1	4	8
8	5	2	4	7	1	6	9	3
4	1	6	9	3	8	2	5	7

58

8	9	3	6	2	1	7	4	5
6	1	5	4	8	7	2	3	9
2	7	4	9	3	5	6	1	8
3	2	7	5	1	6	9	8	4
5	8	6	3	4	9	1	7	2
1	4	9	8	7	2	3	5	6
4	6	2	1	5	3	8	9	7
9	5	1	7	6	8	4	2	3
7	3	8	2	9	4	5	6	1

59

8	7	2	1	5	9	4	6	3
4	1	6	3	2	8	7	9	5
3	5	9	7	6	4	2	1	8
9	6	1	2	7	3	5	8	4
7	4	8	9	1	5	3	2	6
2	3	5	8	4	6	9	7	1
1	8	3	4	9	7	6	5	2
5	9	4	6	8	2	1	3	7
6	2	7	5	3	1	8	4	9

60

6	7	5	3	1	2	4	8	9
4	3	9	8	5	6	7	1	2
8	2	1	9	7	4	6	5	3
3	6	7	4	2	8	1	9	5
9	1	2	5	6	7	3	4	8
5	8	4	1	9	3	2	6	7
1	4	3	7	8	9	5	2	6
2	5	8	6	3	1	9	7	4
7	9	6	2	4	5	8	3	1

61

4	2	1	3	5	9	8	7	6
3	7	6	4	8	1	9	5	2
9	5	8	6	7	2	3	4	1
2	6	4	1	9	8	7	3	5
1	8	5	7	3	4	2	6	9
7	9	3	5	2	6	4	1	8
5	4	2	9	6	7	1	8	3
8	3	7	2	1	5	6	9	4
6	1	9	8	4	3	5	2	7

62

5	7	1	2	4	9	6	3	8
9	4	8	7	3	6	1	5	2
2	3	6	8	5	1	9	7	4
4	1	3	6	9	5	8	2	7
8	6	5	4	2	7	3	9	1
7	9	2	3	1	8	5	4	6
6	8	9	5	7	4	2	1	3
3	5	7	1	8	2	4	6	9
1	2	4	9	6	3	7	8	5

63

9	7	6	1	5	3	2	8	4
4	2	5	8	7	6	9	3	1
1	3	8	2	4	9	7	6	5
5	1	3	9	8	2	4	7	6
6	9	7	3	1	4	5	2	8
2	8	4	7	6	5	3	1	9
3	4	9	6	2	1	8	5	7
8	5	1	4	3	7	6	9	2
7	6	2	5	9	8	1	4	3

64

9	1	5	3	2	6	7	8	4
7	3	4	8	9	1	5	2	6
2	8	6	4	5	7	3	9	1
3	7	1	5	4	2	9	6	8
5	6	9	7	1	8	4	3	2
8	4	2	9	6	3	1	7	5
4	2	3	6	7	5	8	1	9
1	9	7	2	8	4	6	5	3
6	5	8	1	3	9	2	4	7

65

2	6	8	4	9	5	1	3	7
3	5	4	7	1	8	2	6	9
1	9	7	2	3	6	5	4	8
6	4	1	9	8	7	3	5	2
7	3	9	5	6	2	4	8	1
5	8	2	1	4	3	9	7	6
9	1	6	3	7	4	8	2	5
4	7	5	8	2	9	6	1	3
8	2	3	6	5	1	7	9	4

66

9	2	4	8	3	1	6	7	5
1	3	8	7	6	5	4	9	2
6	5	7	9	4	2	1	3	8
4	6	3	5	1	9	8	2	7
7	9	1	3	2	8	5	4	6
2	8	5	4	7	6	3	1	9
5	1	9	2	8	4	7	6	3
8	7	6	1	9	3	2	5	4
3	4	2	6	5	7	9	8	1

67

3	4	1	7	9	6	5	2	8
9	6	2	5	3	8	7	4	1
8	5	7	2	1	4	6	9	3
2	7	4	3	8	1	9	5	6
6	9	8	4	5	2	3	1	7
1	3	5	9	6	7	4	8	2
7	2	9	8	4	3	1	6	5
4	8	6	1	7	5	2	3	9
5	1	3	6	2	9	8	7	4

68

1	8	3	6	7	5	4	9	2
4	7	9	1	3	2	5	8	6
6	5	2	9	4	8	7	3	1
2	1	8	7	9	6	3	4	5
3	9	6	2	5	4	1	7	8
5	4	7	3	8	1	2	6	9
9	2	4	5	6	3	8	1	7
7	3	5	8	1	9	6	2	4
8	6	1	4	2	7	9	5	3

69

7	9	3	4	2	1	5	6	8
5	8	1	6	9	3	7	4	2
6	2	4	7	5	8	9	1	3
9	3	6	8	4	7	2	5	1
1	4	7	5	3	2	8	9	6
8	5	2	9	1	6	4	3	7
3	6	5	2	7	9	1	8	4
2	1	9	3	8	4	6	7	5
4	7	8	1	6	5	3	2	9

70

5	4	1	2	8	3	7	9	6
2	3	9	6	4	7	8	1	5
7	6	8	9	5	1	2	4	3
9	5	3	4	1	8	6	7	2
4	7	6	3	2	9	5	8	1
8	1	2	5	7	6	4	3	9
3	9	5	7	6	4	1	2	8
1	2	7	8	9	5	3	6	4
6	8	4	1	3	2	9	5	7

71

8	7	2	6	1	3	9	4	5
4	1	3	7	5	9	8	2	6
9	6	5	8	2	4	7	1	3
3	4	6	9	7	2	5	8	1
2	8	1	4	6	5	3	7	9
7	5	9	1	3	8	2	6	4
5	3	8	2	4	6	1	9	7
1	9	4	3	8	7	6	5	2
6	2	7	5	9	1	4	3	8

72

1	6	5	4	9	2	8	3	7
3	4	8	7	6	1	5	9	2
2	7	9	3	5	8	1	6	4
8	1	7	5	2	6	9	4	3
6	9	2	1	3	4	7	5	8
4	5	3	9	8	7	6	2	1
5	3	1	8	4	9	2	7	6
7	2	4	6	1	5	3	8	9
9	8	6	2	7	3	4	1	5

73

2	9	4	5	3	1	6	8	7
1	6	8	7	9	4	2	3	5
7	5	3	2	8	6	9	1	4
6	8	2	4	7	5	1	9	3
9	3	7	1	6	8	4	5	2
4	1	5	3	2	9	8	7	6
5	4	6	9	1	7	3	2	8
3	7	1	8	4	2	5	6	9
8	2	9	6	5	3	7	4	1

74

4	1	7	8	2	3	5	6	9
6	8	2	5	4	9	7	1	3
9	5	3	6	1	7	4	2	8
7	2	8	9	5	4	6	3	1
3	9	1	7	6	2	8	5	4
5	6	4	1	3	8	9	7	2
8	7	5	2	9	1	3	4	6
2	4	6	3	8	5	1	9	7
1	3	9	4	7	6	2	8	5

75

1	9	6	5	7	4	8	3	2
8	5	2	3	6	1	7	9	4
4	7	3	2	9	8	1	5	6
5	2	9	6	8	7	3	4	1
7	1	8	4	3	5	6	2	9
6	3	4	9	1	2	5	7	8
3	4	5	8	2	6	9	1	7
9	6	1	7	4	3	2	8	5
2	8	7	1	5	9	4	6	3

76

4	3	1	9	8	5	2	7	6
6	5	2	7	4	1	3	8	9
8	9	7	6	3	2	1	5	4
9	6	3	1	7	8	5	4	2
7	2	8	4	5	9	6	1	3
1	4	5	2	6	3	8	9	7
3	7	4	8	1	6	9	2	5
5	8	9	3	2	4	7	6	1
2	1	6	5	9	7	4	3	8

77

3	1	8	6	5	7	4	9	2
7	9	2	3	4	1	8	6	5
5	6	4	9	8	2	7	3	1
2	7	9	5	3	6	1	4	8
8	3	6	1	7	4	2	5	9
4	5	1	2	9	8	3	7	6
1	4	7	8	6	9	5	2	3
6	8	5	7	2	3	9	1	4
9	2	3	4	1	5	6	8	7

78

3	9	5	8	1	4	7	2	6
6	4	8	2	7	5	3	9	1
7	1	2	9	6	3	8	4	5
2	6	9	7	4	1	5	3	8
8	5	7	6	3	2	4	1	9
1	3	4	5	9	8	6	7	2
9	7	1	4	5	6	2	8	3
5	2	3	1	8	7	9	6	4
4	8	6	3	2	9	1	5	7

79

2	5	3	6	4	1	9	7	8
8	4	1	7	9	5	6	2	3
9	6	7	3	8	2	5	4	1
3	1	2	9	5	8	7	6	4
4	7	5	1	6	3	2	8	9
6	9	8	4	2	7	1	3	5
7	8	4	5	1	6	3	9	2
5	2	6	8	3	9	4	1	7
1	3	9	2	7	4	8	5	6

80

1	9	3	6	5	2	4	8	7
8	2	7	4	9	1	5	3	6
4	6	5	3	8	7	9	2	1
2	4	6	5	1	8	3	7	9
7	1	8	9	2	3	6	4	5
3	5	9	7	6	4	8	1	2
5	8	1	2	4	6	7	9	3
6	3	2	8	7	9	1	5	4
9	7	4	1	3	5	2	6	8

81

5	3	4	7	8	1	9	2	6
6	9	8	5	2	3	7	1	4
7	1	2	6	4	9	8	3	5
9	6	5	1	3	8	2	4	7
8	4	1	2	5	7	3	6	9
2	7	3	4	9	6	1	5	8
1	5	7	8	6	2	4	9	3
3	8	6	9	1	4	5	7	2
4	2	9	3	7	5	6	8	1

82

2	1	3	8	9	5	7	6	4
7	8	6	2	4	3	5	1	9
9	5	4	6	7	1	3	2	8
4	2	9	7	3	6	8	5	1
5	7	8	9	1	2	6	4	3
3	6	1	4	5	8	9	7	2
1	9	7	3	6	4	2	8	5
8	3	5	1	2	7	4	9	6
6	4	2	5	8	9	1	3	7

83

2	3	5	8	7	9	1	6	4
1	6	9	5	3	4	7	2	8
7	8	4	1	2	6	3	9	5
4	9	2	3	6	5	8	7	1
3	1	7	4	9	8	6	5	2
8	5	6	2	1	7	4	3	9
6	4	3	9	5	1	2	8	7
9	7	1	6	8	2	5	4	3
5	2	8	7	4	3	9	1	6

84

3	2	8	7	4	6	5	9	1
7	6	1	5	9	8	2	4	3
5	4	9	2	3	1	7	8	6
2	1	3	9	8	4	6	5	7
9	5	6	1	2	7	8	3	4
4	8	7	6	5	3	9	1	2
6	9	5	3	1	2	4	7	8
1	7	4	8	6	5	3	2	9
8	3	2	4	7	9	1	6	5

85

5	9	3	1	2	8	6	4	7
4	7	1	9	6	3	8	2	5
8	2	6	5	4	7	9	3	1
7	3	2	6	5	9	4	1	8
6	8	9	3	1	4	5	7	2
1	5	4	7	8	2	3	9	6
3	4	8	2	7	6	1	5	9
2	6	5	4	9	1	7	8	3
9	1	7	8	3	5	2	6	4

86

7	2	1	3	4	6	8	5	9
6	3	5	8	9	1	4	7	2
9	8	4	2	7	5	1	3	6
2	4	9	7	6	8	5	1	3
8	1	7	9	5	3	6	2	4
3	5	6	1	2	4	9	8	7
5	7	3	6	1	9	2	4	8
1	9	2	4	8	7	3	6	5
4	6	8	5	3	2	7	9	1

87

7	3	6	8	2	5	9	4	1
4	1	8	9	7	3	2	5	6
2	5	9	4	1	6	7	8	3
9	6	7	5	3	4	8	1	2
5	2	4	7	8	1	6	3	9
1	8	3	6	9	2	4	7	5
8	9	1	2	5	7	3	6	4
3	4	2	1	6	8	5	9	7
6	7	5	3	4	9	1	2	8

88

3	9	5	8	6	4	1	7	2
2	7	4	1	5	3	6	8	9
1	6	8	2	9	7	5	4	3
6	2	3	4	7	5	8	9	1
9	8	7	3	1	6	4	2	5
5	4	1	9	8	2	7	3	6
8	5	2	7	3	1	9	6	4
7	3	6	5	4	9	2	1	8
4	1	9	6	2	8	3	5	7

89

6	4	9	5	3	8	2	7	1
1	3	5	2	7	6	4	8	9
7	2	8	4	1	9	5	3	6
3	5	4	8	9	1	6	2	7
8	1	7	6	2	5	9	4	3
2	9	6	7	4	3	1	5	8
4	6	3	9	5	7	8	1	2
9	7	2	1	8	4	3	6	5
5	8	1	3	6	2	7	9	4

90

3	5	4	6	2	1	8	9	7
1	9	6	4	8	7	3	2	5
2	7	8	5	9	3	6	4	1
7	4	5	9	3	8	1	6	2
9	6	3	2	1	5	4	7	8
8	2	1	7	4	6	5	3	9
6	8	2	1	7	4	9	5	3
4	3	7	8	5	9	2	1	6
5	1	9	3	6	2	7	8	4

91

9	5	1	3	4	6	7	8	2
3	4	6	7	8	2	5	9	1
7	2	8	5	1	9	6	3	4
6	8	7	1	5	4	9	2	3
2	1	9	8	6	3	4	5	7
4	3	5	9	2	7	8	1	6
8	6	4	2	9	1	3	7	5
5	7	2	6	3	8	1	4	9
1	9	3	4	7	5	2	6	8

92

4	3	8	2	1	5	7	9	6
2	7	9	3	4	6	5	1	8
5	6	1	8	7	9	3	4	2
7	5	4	9	3	2	6	8	1
6	8	2	4	5	1	9	7	3
9	1	3	7	6	8	4	2	5
8	4	7	6	2	3	1	5	9
3	9	5	1	8	4	2	6	7
1	2	6	5	9	7	8	3	4

93

8	2	4	7	3	9	1	6	5
6	3	9	1	5	8	7	4	2
1	7	5	6	4	2	9	3	8
5	6	2	9	7	4	3	8	1
3	4	8	2	6	1	5	9	7
7	9	1	5	8	3	4	2	6
9	8	7	3	1	6	2	5	4
4	1	3	8	2	5	6	7	9
2	5	6	4	9	7	8	1	3

94

4	7	9	2	1	3	6	8	5
3	5	2	6	8	9	1	7	4
1	8	6	5	4	7	9	3	2
6	1	4	8	2	5	3	9	7
9	3	8	7	6	4	5	2	1
5	2	7	9	3	1	4	6	8
7	6	1	4	9	8	2	5	3
8	9	3	1	5	2	7	4	6
2	4	5	3	7	6	8	1	9

95

7	2	3	6	9	5	8	4	1
6	9	5	1	4	8	3	7	2
1	4	8	3	2	7	6	9	5
8	1	4	9	7	6	5	2	3
9	5	7	2	8	3	1	6	4
3	6	2	5	1	4	7	8	9
4	8	9	7	5	1	2	3	6
5	7	6	4	3	2	9	1	8
2	3	1	8	6	9	4	5	7

96

7	1	2	9	4	5	3	6	8
9	8	6	2	7	3	5	4	1
5	3	4	6	8	1	9	2	7
6	4	5	3	1	9	8	7	2
3	7	1	8	2	4	6	5	9
2	9	8	7	5	6	1	3	4
4	2	3	5	9	8	7	1	6
1	6	9	4	3	7	2	8	5
8	5	7	1	6	2	4	9	3

97

8	4	5	3	7	9	6	1	2
1	7	9	6	2	4	8	3	5
2	3	6	5	8	1	7	9	4
6	9	4	8	1	2	3	5	7
5	8	3	7	4	6	1	2	9
7	1	2	9	5	3	4	8	6
3	5	7	4	9	8	2	6	1
4	2	8	1	6	5	9	7	3
9	6	1	2	3	7	5	4	8

98

5	1	2	8	9	4	6	7	3
3	9	7	2	1	6	8	5	4
4	8	6	7	5	3	1	2	9
8	3	5	1	2	9	7	4	6
9	6	1	4	7	5	3	8	2
7	2	4	3	6	8	9	1	5
2	5	8	9	3	7	4	6	1
6	7	9	5	4	1	2	3	8
1	4	3	6	8	2	5	9	7

99

7	8	2	9	6	4	3	5	1
1	4	9	5	2	3	6	8	7
5	6	3	7	1	8	2	4	9
4	2	6	3	9	1	8	7	5
8	9	1	6	5	7	4	3	2
3	5	7	4	8	2	9	1	6
2	3	5	1	4	9	7	6	8
9	1	4	8	7	6	5	2	3
6	7	8	2	3	5	1	9	4

100

5	3	7	1	6	4	9	8	2
8	2	6	5	3	9	1	7	4
1	4	9	2	8	7	3	5	6
6	7	8	3	4	1	5	2	9
3	9	1	6	2	5	7	4	8
4	5	2	9	7	8	6	1	3
9	8	4	7	1	6	2	3	5
7	6	3	4	5	2	8	9	1
2	1	5	8	9	3	4	6	7

101

2	8	3	5	7	6	1	4	9
7	1	9	8	4	3	2	5	6
4	5	6	2	1	9	3	8	7
3	6	1	9	8	5	7	2	4
9	4	8	3	2	7	6	1	5
5	2	7	4	6	1	8	9	3
1	9	5	7	3	8	4	6	2
8	3	2	6	9	4	5	7	1
6	7	4	1	5	2	9	3	8

102

5	7	9	1	2	6	3	8	4
3	4	2	9	5	8	6	1	7
1	6	8	7	3	4	5	2	9
6	5	1	4	8	9	7	3	2
2	3	7	6	1	5	4	9	8
8	9	4	3	7	2	1	5	6
7	2	5	8	4	1	9	6	3
9	8	3	5	6	7	2	4	1
4	1	6	2	9	3	8	7	5

103

2	4	5	9	8	7	6	1	3
8	6	7	3	1	5	4	2	9
3	9	1	2	6	4	7	5	8
9	2	4	1	5	3	8	7	6
5	7	8	6	4	2	9	3	1
1	3	6	7	9	8	2	4	5
7	5	3	8	2	9	1	6	4
4	1	9	5	7	6	3	8	2
6	8	2	4	3	1	5	9	7

104

3	6	1	8	5	7	2	9	4
5	2	9	3	4	1	8	7	6
4	8	7	2	9	6	5	3	1
9	3	5	4	8	2	6	1	7
7	1	8	9	6	5	4	2	3
6	4	2	7	1	3	9	8	5
2	5	6	1	7	8	3	4	9
8	7	4	5	3	9	1	6	2
1	9	3	6	2	4	7	5	8

105

3	7	1	2	8	5	9	6	4
8	9	2	6	3	4	1	5	7
6	5	4	7	1	9	8	2	3
1	3	6	4	7	2	5	9	8
7	8	5	9	6	1	4	3	2
4	2	9	8	5	3	7	1	6
5	1	8	3	2	7	6	4	9
9	6	3	5	4	8	2	7	1
2	4	7	1	9	6	3	8	5

106

6	2	3	5	8	1	7	9	4
4	8	5	7	9	3	1	6	2
9	7	1	4	2	6	8	3	5
1	6	8	2	5	4	9	7	3
5	4	9	3	1	7	2	8	6
2	3	7	8	6	9	5	4	1
7	1	4	9	3	2	6	5	8
8	9	6	1	4	5	3	2	7
3	5	2	6	7	8	4	1	9

107

2	8	9	1	3	7	4	6	5
5	1	4	6	8	9	7	2	3
3	7	6	4	2	5	8	9	1
8	3	1	5	7	2	6	4	9
6	5	2	9	1	4	3	8	7
9	4	7	8	6	3	5	1	2
4	9	3	2	5	6	1	7	8
1	6	5	7	9	8	2	3	4
7	2	8	3	4	1	9	5	6

108

7	4	6	9	3	5	2	8	1
1	8	3	4	2	7	5	9	6
2	9	5	6	8	1	7	4	3
3	5	9	7	1	4	8	6	2
8	1	4	2	5	6	9	3	7
6	2	7	8	9	3	4	1	5
9	7	1	5	6	8	3	2	4
5	6	2	3	4	9	1	7	8
4	3	8	1	7	2	6	5	9

109

6	4	3	9	7	1	8	2	5
7	5	9	3	2	8	1	4	6
8	1	2	4	6	5	7	3	9
5	7	6	2	1	9	4	8	3
2	9	1	8	3	4	5	6	7
4	3	8	6	5	7	2	9	1
3	8	5	1	9	2	6	7	4
9	2	7	5	4	6	3	1	8
1	6	4	7	8	3	9	5	2

110

2	9	6	5	7	4	8	1	3
3	8	5	9	6	1	4	2	7
7	4	1	3	2	8	5	6	9
1	2	8	6	5	3	7	9	4
6	3	7	8	4	9	2	5	1
4	5	9	7	1	2	6	3	8
5	6	3	4	9	7	1	8	2
8	1	4	2	3	5	9	7	6
9	7	2	1	8	6	3	4	5

111

4	7	5	3	2	8	9	6	1
2	9	1	5	6	7	3	8	4
8	3	6	1	9	4	5	7	2
9	2	7	8	4	3	1	5	6
6	4	3	2	5	1	7	9	8
5	1	8	6	7	9	4	2	3
1	5	4	9	8	2	6	3	7
3	8	9	7	1	6	2	4	5
7	6	2	4	3	5	8	1	9

112

9	7	4	5	1	6	2	8	3
8	5	6	7	3	2	1	9	4
1	2	3	8	9	4	6	7	5
7	1	5	2	8	3	4	6	9
3	9	2	4	6	1	7	5	8
4	6	8	9	7	5	3	2	1
2	4	1	6	5	8	9	3	7
5	3	7	1	2	9	8	4	6
6	8	9	3	4	7	5	1	2

113

3	9	2	8	7	4	1	6	5
8	1	6	9	2	5	7	3	4
4	7	5	6	3	1	8	9	2
6	2	8	3	1	7	4	5	9
9	5	3	4	6	8	2	7	1
7	4	1	5	9	2	6	8	3
1	3	9	2	8	6	5	4	7
5	6	7	1	4	3	9	2	8
2	8	4	7	5	9	3	1	6

114

7	1	4	9	2	3	6	5	8
3	6	5	7	8	4	2	1	9
8	9	2	6	5	1	3	4	7
1	3	8	4	6	2	9	7	5
6	2	9	5	1	7	8	3	4
4	5	7	8	3	9	1	2	6
5	4	3	1	9	8	7	6	2
2	8	6	3	7	5	4	9	1
9	7	1	2	4	6	5	8	3

115

6	5	3	1	9	7	4	2	8
4	1	7	5	8	2	6	9	3
2	9	8	4	6	3	7	5	1
8	6	4	2	1	9	3	7	5
3	7	5	6	4	8	9	1	2
9	2	1	7	3	5	8	4	6
1	4	9	3	2	6	5	8	7
5	8	6	9	7	1	2	3	4
7	3	2	8	5	4	1	6	9

116

8	6	5	7	3	9	2	4	1
4	7	3	2	5	1	6	9	8
2	1	9	8	6	4	3	7	5
9	3	1	4	7	5	8	2	6
5	8	4	3	2	6	7	1	9
7	2	6	1	9	8	4	5	3
6	4	2	9	1	3	5	8	7
3	9	7	5	8	2	1	6	4
1	5	8	6	4	7	9	3	2

117

6	5	8	3	4	7	1	2	9
1	2	3	9	5	6	7	8	4
7	4	9	1	2	8	3	5	6
5	8	1	6	3	4	9	7	2
9	3	4	5	7	2	8	6	1
2	7	6	8	9	1	5	4	3
8	1	5	2	6	3	4	9	7
3	6	7	4	8	9	2	1	5
4	9	2	7	1	5	6	3	8

118

8	3	2	5	1	4	6	7	9
5	6	4	7	3	9	2	1	8
7	1	9	6	8	2	3	4	5
4	2	6	9	5	1	8	3	7
1	8	7	3	4	6	9	5	2
3	9	5	2	7	8	1	6	4
6	4	1	8	2	7	5	9	3
9	5	8	4	6	3	7	2	1
2	7	3	1	9	5	4	8	6

119

1	2	8	5	3	4	9	7	6
7	5	6	9	8	1	4	3	2
9	3	4	6	2	7	8	5	1
2	4	1	3	6	5	7	9	8
8	7	5	1	9	2	6	4	3
3	6	9	7	4	8	1	2	5
4	8	7	2	5	6	3	1	9
6	9	2	4	1	3	5	8	7
5	1	3	8	7	9	2	6	4

120

7	1	4	3	6	5	9	8	2
6	3	9	7	2	8	5	1	4
2	5	8	4	1	9	7	6	3
3	2	5	1	8	7	4	9	6
4	6	1	9	5	2	8	3	7
8	9	7	6	3	4	2	5	1
1	4	2	5	9	3	6	7	8
5	8	3	2	7	6	1	4	9
9	7	6	8	4	1	3	2	5